T0276205

Nuclear Power

The Power Generation Series

Paul Breeze—Coal-Fired Generation, ISBN 13: 9780128040065

Paul Breeze—Gas-Turbine Fired Generation, ISBN 13: 9780128040058

Paul Breeze—Solar Power Generation, ISBN 13: 9780128040041

Paul Breeze—Wind Power Generation, ISBN 13: 9780128040386

Paul Breeze—Fuel Cells, ISBN 13: 978-0-08-101039-6

Paul Breeze—Energy from Waste, ISBN 13: 978-0-08-101042-6

Paul Breeze—Nuclear Power, ISBN 13: 978-0-08-101043-3

Paul Breeze—Electricity Generation and the Environment,
ISBN 13: 978-0-08-101044-0

Nuclear Power

Paul Breeze

AMSTERDAM • BOSTON • HEIDELBERG • LONDON
NEW YORK • OXFORD • PARIS • SAN DIEGO
SAN FRANCISCO • SINGAPORE • SYDNEY • TOKYO
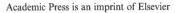
Academic Press is an imprint of Elsevier

Academic Press is an imprint of Elsevier
125 London Wall, London EC2Y 5AS, United Kingdom
525 B Street, Suite 1800, San Diego, CA 92101-4495, United States
50 Hampshire Street, 5th Floor, Cambridge, MA 02139, United States
The Boulevard, Langford Lane, Kidlington, Oxford OX5 1GB, United Kingdom

British Library Cataloguing-in-Publication Data
A catalogue record for this book is available from the British Library

Library of Congress Cataloging-in-Publication Data
A catalog record for this book is available from the Library of Congress

ISBN: 978-0-08-101043-3

For Information on all Academic Press publications
visit our website at https://www.elsevier.com

Working together
to grow libraries in
developing countries

www.elsevier.com • www.bookaid.org

Publisher: Joe Hayton
Acquisition Editor: Lisa Reading
Editorial Project Manager: Maria Convey
Production Project Manager: Mohana Natarajan

Typeset by MPS Limited, Chennai, India

CONTENTS

CHAPTER *1*

An Introduction to Nuclear Power

Nuclear power is the most controversial of all forms of electricity generation. Evaluating its importance involves weighing political, strategic and, often, emotional considerations alongside the more usual technical, economic, and environmental factors that form the core elements of any power technology debate. To complicate the issue further, partisan views are frequently advanced so that balanced, objective decision-making can be difficult, if not impossible.

From a technological point of view, nuclear power offers a tested method of generating electrical power based on exploitation of the energy released during controlled nuclear reactions. However the byproducts of these nuclear processes are a range of extremely toxic radioactive materials, and the nuclear industry has still to develop widely accepted ways of disposing of this waste in a secure manner. An additional, complicating issue is the ability of nuclear technology to be exploited for weapons production as well as power; the overlap between the two makes it very difficult to separate one from the other.

The other great cause of anxiety is the possibility of a nuclear accident. The consequences of an accident at a nuclear power plant that leads to the release of radioactivity into the environment are potentially much more far-reaching that would be the case with an accident at any other type of power plant. This was dramatically demonstrated by the incident at the Fukushima Daiichi nuclear power plant in Japan following a massive undersea earthquake and tsunami in 2011. Both this and the earlier Chernobyl nuclear plant failure have seriously damaged popular confidence in nuclear power and both have had an influence on government policy in several countries, notably Germany, where nuclear power is being phased out as a result of the 2011 accident.

If these issues weigh negatively in the balance, on the positive side nuclear power does offer a carbon-free means of generating electricity and in a world where the control of carbon emissions has become a major requirement if global warming is to be controlled, nuclear technology potentially offers a part of the solution. The International Energy Agency has identified nuclear power as one of the key technologies that will be needed if the globally agreed target of trying to limit global warming to 2°C is to be achieved.[1] Whether the organization's optimistic forecasts for nuclear growth are realistic, given both the negative effect of the 2011 accident and the fall in the cost of power from solar and onshore wind power, which has undercut wholesale power prices in Europe, remains to be seen.

The modern nuclear power industry was born out of the nuclear weapons programs that led to the development of the atom bomb during the 1940s. While weaponry was the aim of the research during World War II, power generation quickly became a secondary, peaceful product of the work and programs were started in both the United States and Russia. Electricity production from nuclear power began on a very small scale at an experimental reactor in the United States in 1951 while in Russia a 5-MW plant designed to generate electricity began operating in 1954. This was followed, in 1956, by a 50-MW dual purpose reactor[2] at Calder Hall, in the United Kingdom, while the French began a nuclear program the same year at Marcoule. The first fully commercial nuclear power plant was built at Shippingport, Pennsylvania, in the United States, and started operating in 1957.

From the late 1950s nuclear generating capacity began to grow rapidly with commercial power plant construction programs in the United States, Russia, the United Kingdom, France, and Canada. Other countries such as Germany, Sweden, and Japan also began to invest in the technology. By the beginning of the 1970s, nuclear power was seen as the great hope for the future of electricity generation. In the United States alone, power companies had ordered over 200 nuclear reactors.

Even as the nuclear power industry was booming, there were concerns about nuclear safety and the consequences of an accident. These were brought into sharp relief by an accident at the Three Mile Island nuclear power plant in the United States in 1979, followed by the catastrophic failure at the Chernobyl reactor in Ukraine in 1986. The reaction to these accidents was a tightening of regulatory control of nuclear power, the introduction of new safety measures, and a loss

[1]Technology Roadmap: Nuclear Power, International Energy Agency and Nuclear Energy Agency, 2015.
[2]The Magnox reactor could produce both power and plutonium.

of public confidence in the technology. The tougher safety regulations meant that the cost of nuclear power soared and in the west the construction of new nuclear power plants was virtually abandoned. It slowed elsewhere too so that during the 1990s and early years of the current century new nuclear plant construction was rare.

Meanwhile the industry was hoping for a renaissance. Interest in nuclear power in China provided one ray of hope and at the same time all the major nuclear power plant manufacturers began to develop a new generation of safer reactors. By 2010 it appeared that a renaissance had begun and orders started to roll once more. However the Fukushima accident in 2011 halted this new blossoming and once again the nuclear power industry is having to face a bleak future.

With the absence of new orders, the main support for the industry is coming from the life extension of existing power stations and reactors. In many countries life extension to 60 years is either being contemplated or taking place, although some of the older nuclear technologies such as the United Kingdom's advanced gas reactors and the Russian RBMK reactors are being phased out completely.

While energy production based on nuclear fission reactions (the splitting of atoms to release energy) may have a limited future, there is another nuclear technology based on nuclear fusion (the fusing of atoms with the release of energy). This technology, too, has its origins in the development of weapons, in this case the hydrogen bomb. Experiments to exploit fusion were started in the 1930s and resumed after an interlude in the 1940s. However, fusion has proved much more difficult to exploit than fission and no commercial scale fusion reactor has been built. Development work continues and hope remains that it can provide an alternative source of energy by the end of the century, if not before.

THE HISTORY OF NUCLEAR POWER

The development of nuclear energy and of nuclear weapons relies on an understanding of atomic structure and of the stability of atomic nuclei. The idea that matter might be composed of atoms was first put forward by Greek philosophers and the name atom is derived from the Greek word *atomos*, which means indivisible. This concept, which became known as atomism, was taken up by a range of ancient religions and philosophies but in the 17th century it began to develop into a more recognizably modern theory of matter through the work of people like René Descartes and Robert Hooke and their speculations about a mechanical universe. These ideas were also taken up by chemists, particularly with the work of Robert Boyle and this gradually fed during the 18th century into more complex concepts about the nature of chemical reactions.

The empirical basis for atomic theory was first laid out by John Dalton in the early 19th century and although it remained controversial throughout the century, theoretical and experimental evidence to support it grew. By the end of the century, the New Zealand scientist Ernest Rutherford was developing his "disintegration theory," which proposed that radioactivity, which was by then a well-known phenomenon, was the result of atomic processes. In 1904 he speculated that if the rate of disintegration of these radio elements could be controlled, an enormous amount of energy could be obtained for a small amount of matter.[3] This idea was formalized mathematically by Albert Einstein in this famous equation:

$$E = mc^2$$

which relates mass and energy. Einstein published this equation in a paper in 1905.

Work continued through the early part of the 20th century on the unraveling of the nature of atoms and their structure. However the next milestone in the development of atomic energy came in 1934 when the physicist Enrico Fermi showed that when neutrons were fired at a range of atoms the particles could cause the atoms to split. His experiments indicated that when the element uranium was treated in this way, the result of the fission process was elements that were much lighter than the original uranium. In 1938 German scientists Otto Hahn and Fritz Strassman performed similar experiments and were able to identify the products as elements such as barium with around half the mass of uranium. A colleague, Lise Meitner, who worked in Copenhagen used their work to calculate that there was a small amount of mass missing when all the fragments were added together and she showed that this had emerged as energy, confirming Einstein's theory and also confirming that atomic fission had taken place.

[3]The History of Nuclear Energy, US Department of Energy.

When a uranium atom is struck by a neutron and splits via a fission reaction, it will commonly produce three more neutrons. This led scientists around the world to begin to speculate about the possibility of putting together enough uranium[4] to create a self-sustaining nuclear reaction, a chain reaction, during which the neutrons released by uranium fission could produce additional, multiple fission reactions. Since each fission reaction also released an enormous amount of energy, this concept had potential both as a way of releasing a massive amount of energy, explosively, in a very short space of time and of providing a controlled release of energy in an energy plant. The exploration of the first of these options through the US Manhattan Project led, in 1945, to the development of the first atomic bomb. Alongside the atomic weapons development, the idea of a uranium reactor was under consideration as early as 1941. This led, at the end of 1942, to the first nuclear reactor, called Chicago Pile-1 that demonstrated the possibility of nuclear reaction being controlled in order to generate energy.

As part of the weapons program, scientists had been working on nuclear fast reactors—often called breeder reactors—that would produce new nuclear material during their operation. An experimental reactor of this type was developed at the Argonne National Laboratory in the US state of Idaho after World War II and in 1951 became the first nuclear reactor to generate electricity from nuclear energy. In 1953 the US president Dwight Eisenhower put forward Atoms for Peace program to direct US nuclear research toward energy production.

Russia also had a longstanding nuclear research program that was partly directed toward energy production, and in 1954 an existing reactor designed for plutonium production was modified for heat and electricity generation. This AM-1 reactor with a generating capacity of 5 MW became the first nuclear power station. In the United States, meanwhile, development of a pressurized water reactor (PWR) for naval use was underway, leading to the first nuclear powered submarine in 1954. The design of this reactor led to the first large-scale demonstration nuclear power plant, the 60-MW Shippingport reactor in Pennsylvania that started operating in 1957. Following that, the first fully commercial nuclear station, the 250-MW Westinghouse PWR, Yankee Rowe, started operating in 1960. An alternative design, the

---

[4]The amount required is called the critical mass.

boiling water reactor (BWR) was also under development and the first 250-MW station based on this design also started operating in 1960.

In the United Kingdom another design, the gas-cooled Magnox reactor was developed and the first 50-MW dual purpose reactor to this design began operating at Calder Hall in 1956. In France a reactor of similar design started operating in 1956 and commercial plants began to appear in 1959. Russia took slightly longer to develop power plants and it was 1964 before the first two nuclear plants, based on a Russian BWR design started to operate.

In the first decade of the 21st century, while there are still a range of different reactor designs in operation, by far the largest portion of these are now PWRs.

GLOBAL ELECTRICITY PRODUCTION FROM NUCLEAR POWER

From the beginning of the 1970s, the output from the world's nuclear power plants grew steadily as their numbers increased. However the slowdown in construction that began during the 1990s saw the rise in output slow. Total output peaked during the middle of the first decade of the 21st century and since then it has declined. With output static or declining, the proportion of global electric power generated by nuclear power plants has been falling since the middle of the 1990s.

Table 1.1 shows the number of reactors at the end of 2015, broken down by type. There were 442 operating reactors but three of these

Table 1.1 Nuclear Power Reactors in Operation at the End of 2015		
Reactor Type	**Number**	**Proportion of Total (%)**
Pressurized water reactor	283	64.0
Boiling water reactor	78	17.6
Pressurized heavy water reactor	49	11.1
Light water graphite reactor	15	3.4
Advanced gas-cooled reactor	14	3.2
Nuclear fast reactor	3	0.7
Total	442	100.0
Source: *World Nuclear Association.*[5]		

[5]World Nuclear Performance Report 2016, World Nuclear Association.

were experimental fast breeder reactors; the remaining 439 were commercial power reactors. The largest group comprises 283 PWRs and these make up 64.0% of the total. The alternative BWR accounted for 78 further units or 17.6% of all those in operation. There are 15 light water graphite reactors, found only in countries of the former Soviet Union, and 14 advanced gas-cooled reactors, a design unique to the United Kingdom.

Table 1.2 contains figures for the total annual generation from nuclear power plants between 2004 and 2013. Total output in 2004 was 2738 TWh, 15.7% of total global electricity production. Output rose over the next 2 years, reaching 2793 TWh in 2007. Total production continued to hover around 2700 TWh until 2011 when Japanese power plants were shut down following the Fukushima accident. At the end of 2013, total nuclear output was 2478 TWh but with rising global generation, this accounted for only 10.6% of the global total. According to separate figures from the World Nuclear Association, total output in 2015 was 2441 TWh, again around 10% of global production. However figures from the Nuclear Energy Agency indicate that nuclear power in OECD[6] countries accounted for 19.0% of the total electricity production in these countries, nearly twice as high as the global figure.

Table 1.2 Annual Global Electricity Production From Nuclear Power Plants			
Year	Global Nuclear Power Generation (TWh)	Total Global Power Generation (TWh)	Nuclear Production as a Proportion of Annual Production (%)
2004	2738	17,450	15.7
2005	2768	18,239	15.2
2006	2793	18,930	14.8
2007	2719	19,771	13.7
2008	2731	20,181	13.5
2009	2697	20,055	13.4
2010	2756	21,431	12.9
2011	2584	22,126	11.7
2012	2461	22,668	10.9
2013	2478	23,322	10.6
Source: International Energy Agency.[7]			

[6]Organisation for Economic Co-operation and Development.
[7]Key World Statistics 2006–2015.

Table 1.3 Annual Global Nuclear Generating Capacity	
Year	Global Nuclear Generating Capacity (GW)
2004	357
2005	368
2006	369
2007	372
2008	373
2009	371
2010	375
2011	369
2012	373
2013	372
Source: *International Energy Agency.*[8]	

Table 1.3 shows figures for the annual total global nuclear generating capacity between 2004 and 2013. Total capacity in 2004 as 357 MW. Capacity rose during the rest of the decade, peaking at 375 MW in 2010. Generating capacity dipped to 369 MW in 2011 but by 2013 it was again 372 MW.

In Table 1.4 global nuclear electricity output is broken down by country for the top 10 nuclear nations. The country with the largest nuclear production was the United States with 822 TWh. France produced 424 TWh and the Russian Federation 173 TWh. Output in South Korea was 239 TWh, higher than that in China where 112 TWh were produced, while nuclear power plants in Canada generated 103 TWh. Germany produced 97 TWh, Ukraine 83 TWh, the United Kingdom 71 TWh, and Sweden 66 TWh. Production in the rest of the world was 388 TWh. The figures in this table do not include Japan because in 2013 its nuclear power plants were shut down. However in 2010 the last full year of operation of Japan's nuclear fleet, it produced 288 TWh and was the third largest nuclear generator.

Table 1.4 also shows what proportion of each country's electricity is generated by nuclear power plants. The highest proportion is found in France where nearly 75% of the nation's power came from nuclear generation in 2013. Ukraine and Sweden both produced 43% of their electricity from nuclear stations and in South Korea the total was 26%.

[8]Key World Statistics 2006–2015.

Table 1.4 Nuclear Electricity Production by Country in 2013 (TWh)		
Country	Nuclear Power Production (TWh)	Nuclear Production as a Proportion of National Total (%)
United States	822	19.2
France	424	74.7
Russian Federation	173	16.3
South Korea	139	25.8
China	112	2.1
Canada	103	15.8
Germany	97	15.5
Ukraine	83	43.0
United Kingdom	71	19.8
Sweden	66	43.4
Rest of the World	388	7.9
World	2478	10.6
Source: *International Energy Agency.*[9]		

(The production from Japan's plants in 2010 was also 26% of its total.) The United Kingdom produced around 20% of its power from its nuclear stations while in the United States it was 19% and in the Russian Federation 16%. Canada generated 16% of its electricity from nuclear plants, as did Germany, while China's nuclear output accounted for only 2% of its total.

Finally, Table 1.5 shows all the world's nuclear nations, arranged alphabetically, together with their nuclear capacities in 2013. There are 29 in total. As would be expected from the figures above for generation, the United States had the largest installed capacity, 98,990 MW. France, with the second largest fleet had 63,130 MW while Japan had 40,830 MW. Other nations with large nuclear fleets included China with 26,967 MW, Russia with 26,053 MW, and South Korea with 23,017 MW. Canada had 13,553 MW, Ukraine 13,107 MW, and Germany 10,728 MW. All the other countries in the table had less than 10,000 MW of nuclear capacity.

It is notable that several countries with large nuclear fleets have built them to make up for very limited fossil fuel reserves. These countries include Japan, South Korea, and Sweden. Sweden has a

[9]Key World Statistics 2015. The figures are for 2013.

Table 1.5 Installed Nuclear Capacity by Country	
Country	Operating Reactor Capacity (MW)[a]
Argentina	1627
Armenia	376
Belgium	5943
Brazil	1901
Bulgaria	1926
Canada	13,553
China	26,967
Czech Republic	3904
Finland	2741
France	63,130
Germany	10,728
Hungary	1889
India	5302
Japan	40,480
South Korea	23,017
Mexico	1600
Netherlands	485
Pakistan	725
Romania	1310
Russia	26,053
Slovakia	1816
Slovenia	696
South Africa	1830
Spain	7121
Sweden	8849
Switzerland	3333
Ukraine	13,107
United Kingdom	8883
United States	98,990
World Total	384,006

[a]Figures are for operating reactors on May 1, 2016.
Source: World Nuclear Association.

significant hydropower reserve which it also exploits. However, Japan and South Korea have few domestic resources for power generation and both import significant amounts of coal and natural gas in addition to operating nuclear fleets.

CHAPTER 2

Nuclear Fuel and the Nuclear Resource

Nuclear reactors require nuclear fuel in order to function. This fuel is usually uranium although other elements, including plutonium can also be used. Thorium, though not a nuclear fuel itself, can be turned into a suitable isotope of uranium in a nuclear reactor. Thorium is naturally occurring, like uranium, but plutonium is only produced during nuclear reactions so its main source is nuclear reactors.

Uranium is present in most rocks and in seawater and is a relatively common element in the earth's crust with an abundance similar to that of beryllium, molybdenum, tin, arsenic, and germanium. It is found in relatively high concentrations in a few areas and it is these that provide the supply that is used for nuclear power. In addition to the mining of uranium, since the mid-1990s surplus weapons-grade uranium from stockpiles in the former Soviet Union and the United States have also been used to make nuclear power plant fuel.

The known reserves of uranium are considered sufficient to provide fuel for nuclear power plants for at least another century although this may partly depend on whether new nuclear stations are built, and demand increases. If supply were constrained in the future, it would be for economic reasons if uranium becomes too costly to produce. It is unlikely that the uranium in the earth's crust would become exhausted.

Thorium is even more abundant than uranium, with quantities similar to lead and boron present in the earth's crust. It is found in relatively high quantities in India where there is a thorium reactor research program. China is also working on a thorium reactor.

THE NUCLEAR FUEL CYCLE

The production of fuel for nuclear reactors and the handling of spent fuel after it has been removed from a reactor involve a number of industrial processes that taken together are known as the nuclear fuel cycle.

The nuclear fuel cycle

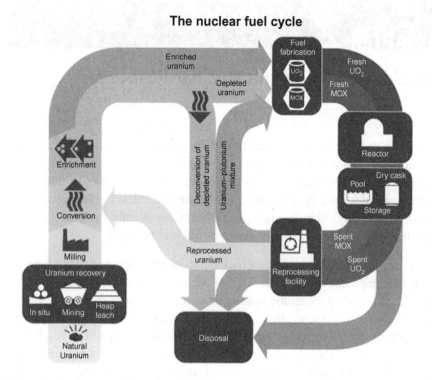

Figure 2.1 Schematic diagram of the nuclear fuel cycle. Source: United States Nuclear Regulatory Commission.

These are shown in Fig. 2.1. The term cycle implies a regenerative loop and it is possible to process used fuel in order to generate new fuel. However, this does not take place in many countries so in most cases "cycle" is a misnomer. Nevertheless the prospect exists, in principle.

The nuclear fuel cycle starts with the mining of uranium-containing ores and the milling of the ore to extract uranium in the form of uranium oxide (U_3O_8). This usually involves the processing of large quantities of relatively low-quality ore, crushing and grinding it in order to release the uranium mineral particles and then capturing the uranium in solution, often with sulfuric acid. The uranium is then extracted from the acid solution to provide a solid oxide (U_3O_8) called yellow cake, which is packaged into drums for shipment to fuel manufacturing facilities.

Some reactors can operate with natural uranium as fuel. For these, the yellow cake can be processed directly into fuel pellets and fuel rods as discussed below. However for most of the world's reactors, naturally occurring uranium is not sufficiently active and it must be enriched so that the proportion of U-235, the main uranium isotope responsible for fission, is increased. Naturally occurring uranium only contains around 0.7% of this isotope.

Uranium enrichment can be carried out in a number of ways but two are common today, gaseous diffusion and gas centrifuge enrichment. In both cases the proportion of U-235 is increased to around 5%. In both too, the yellow cake, U_3O_8, is first converted into uranium hexafluoride, UF_6, before the enrichment process begins. This compound is a solid at ambient temperature but sublimes directly into the gaseous phase at temperatures above 134°C. The enrichment processes are both carried out on gaseous UF_6.

The gaseous diffusion process relies on the slightly different diffusion rates of U-235 and U-238 through a porous membrane to create a stream of gas that is richer in one isotope than the other, as shown schematically in Fig. 2.2. This difference in diffusion rate is tiny for a single membrane so cascades of membranes are required in order the achieve a suitable level of enrichment. Each diffuser in the cascade receives high-pressure UF_6. As this high-pressure gas stream passes through the diffuser and past the membrane, some of the gas diffuses through the membrane into a lower pressure region. The gas in this low-pressure region is slightly enriched with U-235 while the high-pressure flow is slightly depleted. By repeating this process many times, it is possible to achieve any level of enrichment up to 90%, the level required for weapons. Diffusion plants have operated since the middle of the 20th century.

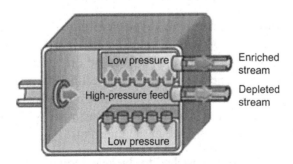

Figure 2.2 Uranium enrichment by gaseous diffusion. Source: US NRC.

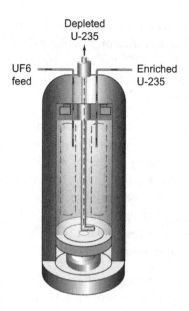

Figure 2.3 Uranium enrichment using a gas centrifuge. Source: Centrus.

Toward the latter part of the 20th century, gas centrifuge enrichment began to take over as the preferred method of uranium enrichment. A single centrifuge is shown schematically in Fig. 2.3. Centrifuges exploit the tiny difference in mass between the two uranium isotopes to achieve a separation. As with the gaseous diffusion process, the separation achieved in each centrifuge is tiny and cascades of centrifuges are required to achieve fuel-level enrichment, or higher. However the centrifuge process is quicker and less energy-intensive than the gaseous diffusion process. This is an advantage from a technical perspective but does make control of nuclear proliferation more difficult.

A third method of enrichment called laser enrichment is also under development. This process uses a tuneable laser that can provide light at a specific frequency, light which will cause one uranium isotope to undergo a chemical reaction while not affecting the other isotopes. A commercial plant based on this technology has been licensed for construction in the United States. Laser enrichment is expected to be more efficient than gas centrifuge enrichment.

Once uranium has been enriched to the level required for nuclear fuel, it must be converted from UF_6 into another uranium oxide, uranium dioxide (UO_2). The UO_2 is then pressed into pellets that are sintered at a high temperature to create tiny, refractory fuel elements. These pellets are loaded into metal fuel rods, ready for insertion into a reactor core where the controlled nuclear reaction takes place. The rods themselves are long, hollow cylinders. They are usually made from zirconium or a zirconium alloy because the element is transparent to neutrons and has good corrosion resistance, especially at high temperatures.

Eventually, when the amount of fissile U-235 in the rods has fallen to a low enough level, these rods of uranium oxide fuel become exhausted and are no longer useful in the reactor. By this stage the nuclear reactions that have taken place have left a range of radioisotopes with short half-lives, so the rods are much more radioactive than when they were loaded into the reactor. Since these reactions release energy, they are hot and continue to generate heat even when removed from the core. Reactor cores are designed so that between 25% and 33% of the fuel rods are removed every year or two, depending upon the specific reactor design.

When the exhausted rods are removed from the reactor, they are placed in water-filled storage tanks alongside the reactor where they cool, both thermally and radioactively. The rods will remain in these storage pools for 2−4 years, longer in some cases, during which time they are sorted and moved around as their level of radioactivity falls.[1]

The fuel rods containing the spent nuclear fuel will eventually be recovered from the storage tanks and will then be treated in one of two ways. The rods might simply be cut up into small pieces and converted into a form that can be permanently stored safely. When spent fuel is handled in this way, there is no fuel cycle. Instead the fuel is usually secured in a vitreous matrix so that it can be placed in a long-term repository. The ideal storage site is considered to be underground in stable rock formations. Unfortunately, there are currently no long-term storage facilities of this type in operation although one has been approved for construction in Finland. Elsewhere, approval to build such repositories has proved difficult to secure.

---

[1]In some countries where there is no fuel reprocessing or storage facility, storage tanks have in the past become the de facto storage facility for spent nuclear fuel. Eventually, however, the storage tanks become full and some other solution has to be found.

The second option is to reprocess the spent fuel to remove the uranium it still contains as well as any plutonium that has been formed during the fission reactions inside the reactor. Both these can then be reused. Uranium can be used to make more enriched uranium and more fuel rods while plutonium and uranium—both in oxide form— can be mixed to create what is known as mixed oxide fuel. This can also be "burnt"[2] in some types of reactor. With this reprocessing, there is a true nuclear fuel cycle since spent fuel is being converted back into usable fuel. However, there is still a significant amount of high level nuclear waste produced too.

In addition to the uranium fuel production process outlined above, there has, since the 1990s, been a further source of enriched uranium, surplus weapons-grade material from the United States and Russian stockpiles. This highly enriched uranium, which can contain up to 90% U-235, can be mixed with natural uranium to create fuel-grade uranium that can be utilized in the same way as normal fuel. According to the World Nuclear Association, this weapons-grade uranium met up to 19% of world reactor demand in 2013. However by 2014 the amount available from this source was falling and it only provided around 9% of the global total.

GLOBAL URANIUM RESOURCES

Uranium is found widely within the earth's crust and is present in most soils, in rock, and in seawater and ground water at very low concentrations. It is known in over 200 mineral forms. The average concentration in the earth's crust is around 2.5 ppm. The element, with atomic number 92 in the periodic table of elements, occurs in several different isotopic forms that vary in the number of neutrons their nucleii contain. The most common isotopes are U-238, U-235, and U-234. Naturally occurring uranium is 99.3% U-238, 0.7% U-235, and a small amount (0.006%) U-234. Uranium is usually considered to be the largest naturally occurring element in the periodic table although elements 93 and 94, neptunium and plutonium do occur in very small quantities in uranium ores as a result of nuclear reactions.

---

[2]The consumption of nuclear fuel in a nuclear power plant core is often referred to as burning, by analogy with fossil fuel plants, even though no combustion process actually takes place.

Although uranium can be recovered from many different sources, it is not normally economical unless there is a relatively high level present. Uranium-containing ores suitable for mining and milling are found in a limited number or regions. The very highest grade ores contain 17–18% uranium (or 200,000 ppm since the quality is often expressed in parts per million). These are only found in two deposits in Canada. More typically, a high-grade ore contains around 2% uranium (20,000 ppm), a low-grade ore contains 0.1% uranium (1000 ppm), while very low-grade ores, such as those found in Namibia, contain only 0.01% uranium (100 ppm).

The value of an ore depends not only upon the concentration of uranium it contains but also upon the ease with which that uranium can be extracted. The size of the global uranium resource is often classified in terms of the amount that is known to exist with an extraction cost below a certain fixed point. The key reference text for uranium supply is commonly known as the "Red Book" and is published every 2 years by the OECD Nuclear Energy Agency and the International Atomic Energy Agency (IAEA). The full title for the latest version of the Red Book is *Uranium 2014: Resources, Production and Demand.*

Figures from this publication for global uranium resources, broken down by country, are shown in Table 2.1 for all those countries with 1% or more of the global resource. The known recoverable resource shown in the table is defined as the Reasonably Assured Resources and the Inferred Resources that are recoverable at a cost of less than US$130 per kgU. The total global amount in the category in 2013 was 5,902,900 tonnes. For the resource that can be recovered for a cost of less than $US260 per kgU, the equivalent amount was 7.096,600 tonnes.

The largest known recoverable deposits of uranium in the world are found in Australia which has 1,706,100 tonnes, or 29% of the global total. Kazakhstan has a further 679,300 tonnes, 12% of the global reserve. Other important deposits are found in the Russian Federation, Canada, Niger, Namibia, and South Africa. As the table shows, there are important uranium deposits in Africa, with significant amounts in Botswana and Tanzania as well as the countries already listed. Asian deposits include those in China and Mongolia. In Central and South America, only Brazil has significant amounts of uranium and the sole European country beside the Russian Federation with a sizable deposit is Ukraine.

Table 2.1 Global Uranium Resources, 2013

Country	Known Recoverable Resource (Tonnes)	Proportion of World Total (%)
Australia	1,706,100	29
Kazakhstan	679,300	12
Russian Federation	505,900	9
Canada	493,900	8
Niger	404,900	7
Namibia	382,800	6
South Africa	338,100	6
Brazil	276,100	5
USA	207,400	4
China	199,100	4
Mongolia	141,500	2
Ukraine	117,700	2
Uzbekistan	91,300	2
Botswana	68,800	1
Tanzania	58,100	1
Jordan	40,000	1
Other	191,900	3
Total	5,902,900	100

Source: OECD-NEA/IAEA.[3]

URANIUM PRODUCTION AND CONSUMPTION

Global uranium production figures for 2012 are shown in Table 2.2. As would be expected, the major producers are primarily the nations with the largest uranium resources. By far the largest producer in 2012 was Kazakhstan, with 21,240 tonnes, 36% of the world total. Canada produced 8998 tonnes, 15% of the total, and Australia a further 7009 tonnnes (12%). These three between them provided 63% of total global production that year. Other important sources include Niger, Namibia, and Malawi in Africa, the Russian Federation and Uzbekistan, Ukraine, the United States, and China. Market conditions toward the middle of the decade have slowed production in some African countries compared to 2014.

Figures in Table 2.3 show the consumption of uranium in 2012, broken down by world regions. The largest regional consumer is

[3]IAEA and OECD-NEA Uranium 2014: Resources, Production and Demand.

Table 2.2 Top 10 Uranium Producers in 2012		
Country	Uranium Production (Tonnes)	Proportion of Global Total (%)
Kazakhstan	21,240	36
Canada	8998	15
Australia	7009	12
Niger	4822	8
Namibia	4653	8
Russian Federation	2862	5
Uzbekistan	2400	4
USA	1667	3
China	1450	2
Malawi	1103	2
Ukraine	1012	2
Others	1600	3
Total	58,816	100

Source: IAEA/OECD-NEA.[4]

Table 2.3 World Uranium Requirement, 2012	
Region	Uranium Requirement (Tonnes)
North America	24,856
European Union	17,235
East Asia	11,180
Non-EU Europe	6635
Middle East, Central, and Southern Asia	875
Central and South America	520
Africa	290
Total	61,600

Source: IAEA/OECD-NEA.[5]

North America, home to the largest national fleet of nuclear units, that of the United States. The uranium requirement in North America in 2012 was 24,856 tonnes. The European Union was also a large consumer with a total requirement of 17,235 tonnes. East Asian demand was for 11,180 tonnes, while in Non-EU Europe there was a requirement for 6636 tonnes. The three other regions in the table each required less than 1000 tonnes in 2012.

[4]Uranium 2014: Resources, Production and Demand, IAEA/OECD-NEA 2014.
[5]Uranium 2014: Resources, Production and Demand, IAEA/OECD-NEA 2014.

Comparing the figures in Tables 2.2 and 2.3, it is clear that the uranium requirement in 2012 exceeded production. As already noted, this shortfall was primarily made up by the use of weapons-grade uranium to make reactor fuel. As the latter is used up, so the demand for production will rise.

THORIUM

The extent of the world's thorium resource has not been mapped as thoroughly as that for uranium but there is thought to be around 6.2 million tonnes of total known and estimated resource according to the Red Book. The most common source is a phosphate mineral, Monazite, which contains an average of around 6% thorium although it can be as high as 12%. The mineral is globally important because it also contains a variety of rare earth elements such as lanthanum and cerium.

The world Monazite resource is estimated to be around 16 million tonnes, of which 12 million tonnes are found in south and east India. Other countries with significant amounts include Brazil, Australia, the United States, Egypt, Turkey, and Venezuela.

Thorium is not itself a nuclear fuel but can be converted in a reactor into U-233, which is useful as a fuel. The exploitation of thorium is therefore more complex that for uranium. There are currently no commercial reactors based around the use of thorium. However, there are experimental programs underway. In particular a company in Norway is exploring the use of fuels containing thorium in existing nuclear power plants.

CHAPTER 3

The Basics of Nuclear Power

Nuclear power reactors release energy in the form of heat from reactions involving the nuclei of atoms. In this context, the reactor is conceptually identical to the boiler in a coal-fired power station, which also releases heat energy from a fuel. In both cases that heat energy is then captured within the boiler walls (in the nuclear plant the boiler is called a steam generator) where water is converted to high-pressure, high-temperature steam and this steam is used to drive a steam turbine generator.

The steam generated by a nuclear power plant never reaches the same temperature and pressure as that in a modern coal-fired boiler; in consequence of this, the steam turbines in nuclear power plants are often much larger than in the fossil fuel plants. There are also design differences resulting from the need to keep all the radioactivity within the confines of the reactor chamber. Nevertheless, apart from the difference in energy source, the layouts of the two types of power station are broadly similar.

The part of a nuclear power plant containing the nuclear reactor and other elements that are exposed to radiation is called the nuclear island. This is normally housed within a large concrete vessel called the containment, a safety-vessel designed to withstand explosions and external impacts. The primary steam generators for a nuclear plant are normally within the containment. However, much of the steam cycle, including the steam turbine generators lie outside the containment. Taken together, these components are called the conventional island. The actual boundary between the nuclear island and the conventional island will depend upon the specific reactor design.

NUCLEAR REACTIONS

The nuclear reactions that generate energy in a nuclear reactor involve the reconfiguring of atomic nuclei. The forces binding atomic nuclei together are incredibly strong and so the amount of energy involved in changing nuclear configurations during a nuclear reaction is large.

The reactions that take place require a rearrangement of the building blocks of atomic nuclei, the protons and neutrons, known collectively as nucleons. In effect, one element is transmuted into another. These processes can be considered to be reactions that create more stable nuclear configurations from less stable ones in the same way as chemical reactions usually create more stable molecules from less stable ones. As with chemical reactions, some of these processes are more or less spontaneous, others initially require an enormous amount of energy to push them to a conclusion, though once started they become self-sustaining. The net result in each case is that the mass of the product nuclei is slightly smaller than that of the starting nuclei. (More precisely, the protons and neutrons in the product nuclei are bound together more strongly than they were in their starting nuclei.) This small mass loss emerges from the reactions as energy which, by Einstein's equation relating the two, leads to a very large amount of energy being released.

All the nuclear power stations operating today generate electricity by utilizing energy released when the nuclei of a large atom such as uranium are split into smaller components, a process called nuclear fission. This reaction can occur spontaneously in nature and can also be triggered relatively easily in certain atomic species—called fissile isotopes—such as uranium-235 in a nuclear reactor. The amount of energy released by uranium fission is enormous. One kilogram of naturally occurring uranium could, in theory, release around 140 GWh of energy. (140 GWh represents the output of a 1000-MW coal-fired plant operating a full power for nearly 6 days.)

There is another type of nuclear reaction, nuclear fusion, which involves the reverse of a fission reaction. In this case small atoms are encouraged to fuse at extraordinarily high temperatures to form larger atoms. Like nuclear fission, fusion releases massive amounts of energy. However, fusion reactions will only take place under extreme conditions and do not occur spontaneously under any conditions found naturally upon the Earth. The fusion of hydrogen atoms is the main source of energy within the Sun and it is a similar reaction that forms the basis for research into nuclear fusion for power generation.

The reason why both fission and fusion can release energy lies in the relative stability of different elements. It turns out that atomic species in the middle of the periodic table of elements, elements such as barium and krypton (these are typical products of uranium fission) are generally more stable than either lighter elements such as hydrogen or heavier elements such as uranium. The nuclei of these more stable atoms are bound together more strongly, and their nuclear components, the protons and the neutrons, require more energy to separate them than those in lighter or heavier nuclei. When less stable nuclei are converted into more stable ones, energy is released. Conversely, energy must be absorbed to turn these more stable nuclei back into less stable ones.

NUCLEAR FISSION

Many large and even some small atoms undergo nuclear fission reactions naturally. One of the isotopes of carbon (isotopes are atoms of a single element with different numbers of neutrons) called carbon-14 behaves in this way. Carbon-14 exists at a constant concentration in natural sources of carbon. Thus living entities that constantly exchange their carbon with the biosphere maintain this constant concentration. However, when they die, the carbon-14 is no longer renewed and it gradually decays. Measuring the residual concentration gives a good estimate of the time since the organism died. It is this property that allows archeologists to use carbon-14 to date ancient artefacts and remains.

Carbon-14 is unstable because of the relative number of protons and neutrons that nucleus contains. Atoms of the most common form of carbon, carbon-12 contain six protons—the positively charged nucleons that confer charge on the nucleus—and six neutrons. (In nuclear reactions this is written $^{12}_{6}C$ to show the numbers of protons and neutrons) A second stable isotope, carbon-13 contains six protons and seven neutrons. Carbon-14, in contrast, contains six protons and eight neutrons. For light elements such as carbon a ratio of protons to neutrons of around 1.1 appears to be the most stable configuration. The ratio for carbon-14, 1.33, is well above this and the nucleus is not stable.

Other atoms can be induced to undergo fission by bombarding them with subatomic particles. One of the isotopes of uranium, the element most widely used in nuclear reactors, behaves in this manner

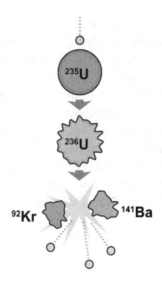

Figure 3.1 Uranium-235 fission. Source: Wikipedia.

(see Fig. 3.1). Naturally occurring uranium is composed primarily of two slightly different isotopes called uranium-235 and uranium-238 (the numbers refer to the sum of protons and neutrons each atom contains).

Most uranium is uranium-238, but 0.7% is uranium-235. Uranium-235 is naturally radioactive with a half-life of 703.8 million years. It decays by releasing an alpha particle[1] to produce thorium. However, it is also fissile which means it is capable of undergoing a fission reaction on absorption of a neutron. The likelihood of reaction depends on the speed of the neutron and it is much more likely to happen with a slow neutron.

When an atom of uranium-235 is struck by a slow neutron, it has a high chance of undergoing a nuclear fission reaction. The most frequent products of this reaction are an atom of krypton, an atom of barium, three more neutrons, and a significant quantity of energy.

$$^{237}_{92}U + n = {}^{140}_{56}Ba + {}^{96}_{36}Kr + 3n + @200 \, meV$$

In theory each of the three neutrons produced during this reaction could strike another atom of uranium-235, causing three additional

---

[1]An alpha particle is a helium nucleus with two protons and two neutrons.

fission reactions. However, this also depends on the quantity of uranium present. If a piece of uranium is too small, then most of the neutrons will escape into the surroundings without ever meeting another uranium-235 nucleus. As the size of the piece of uranium increases, so the likelihood of a neutron from a fission reaction being absorbed by another uranium-235 atom increases. The critical point arrives when the piece of uranium is large enough so that the probability of one of the neutrons from a single fission reaction striking another uranium-235 atom, stimulating a second reaction reaches one. This size is known as the critical mass. Once the critical mass is exceeded and the probability of further fission reactions exceeds one, even by a tiny amount, the result is a rapidly accelerating series of individual fission reactions called a chain reaction. The chain reaction releases an enormous amount of energy in a very short space of time and forms the basis for the atomic bomb.

In fact a critical mass of natural uranium will probably not explode because, as already noted above, the uranium-235 atoms have a high probability of undergoing a fission reaction only when struck by slow moving neutrons. However the neutrons created during the fission process are fast neutrons, too fast to readily induce further fission reactions to take place. They need to be slowed down first. This is crucial to the development of nuclear power.

CONTROLLED NUCLEAR REACTION

In order for uranium fission is to be harnessed in a power station, the nuclear chain reaction must first be tamed. The chain reaction is explosive and dangerous. However it can be managed by carrying away the energy released by the fission reactions, by controlling the number of neutrons within the reactor core, and then by slowing the remaining neutrons so that they can initiate more fission reactions.

An accelerating chain reaction will take place when each fission reaction causes more than one further identical reaction. If the fission of a single uranium-235 atom causes only one identical reaction to take place, the reaction will carry on indefinitely—or at least until the supply of uranium-235 has been used up—without accelerating. But if each fission reaction leads to an average of less than one further reaction, the process will eventually die away naturally. If the number of reactions resulting from each fission reaction can be controlled then, in principle, the nuclear chain reaction can be tamed. A reactor in which each nuclear reaction produces one further nuclear reaction is described as critical. Once the product of each nuclear reaction is more than one additional reaction, the reactor is described as supercritical. Operation must be controlled so that the reactor is just—but barely—supercritical.

To achieve this, the reactor core must contain more than the critical mass of uranium. Otherwise even with the ability to control the number of neutrons, the chain reaction would soon cease because the mass of uranium would rapidly fall below the critical level. Since there is more uranium than is needed to cause a chain reaction, there must also be a way of absorbing and removing neutrons inside the reactor in order to stop a runaway reaction taking place. Finally there must also be a way of slowing the fast neutrons generated by the fission of uranium-235.

Although uranium-235 forms only a small part of natural uranium, it is possible to build a reactor that uses natural uranium for fuel. However, it is easier to build a reactor that uses fuel containing more than the natural amount of uranium-235. Uranium enrichment plants are used to produce reactor fuel that typically has around 3.0−5.0% uranium-235 instead of the 0.7% found naturally. Enriched uranium makes it easier to start and control a sustained nuclear fission reaction.

The uranium is loaded into the reactor core in the form of pellets inside special rods, as described in Chapter 2. In addition to the uranium, the reactor also contains rods made of boron. Boron is capable of absorbing the neutrons generated during the nuclear reaction of uranium-235. If a sufficiently large amount of boron is included within the reactor core, it will absorb and remove the neutrons generated during the fission reaction, stopping the chain reaction from proceeding by keeping the reactor subcritical. The boron rods are moveable and by moving the rods in and out of the reactor core, the number (or flux) of neutrons and hence nuclear process can be controlled.

One further crucial component is needed to make the reactor work, something to slow down the fast neutrons. The neutrons from each uranium-235 fission move too fast to easily stimulate further reactions but they can be slowed by adding a material called a moderator.

Water makes a good moderator and is used in most operating reactors. Graphite also functions well as a moderator and has been used in some reactor designs.

When a uranium fission reaction takes place, the energy it releases emerges as kinetic energy. In other words the products of the fission process carry the energy away as energy of motion; they move extremely fast. (Fast neutrons initially travel at around 3% of the speed of light.) Much of the energy from the fission of uranium-235 is carried away by the three fast neutrons produced during the reaction. These neutrons will dissipate their energy in collision with atoms and molecules within the reactor core. In many reactors this energy is absorbed by the moderator, water. So while the neutrons are slowed, the water within the core becomes hotter. By cycling the water from the reactor core through a heat exchanger, this heat can be extracted and used to raise steam and generate electricity. Extracting the heat also helps maintain the reactor in a stable condition by preventing overheating.

The operation of a nuclear fission reactor is therefore a careful balancing act. As a consequence, a reactor always has the potential to generate a runaway chain reaction. Modern reactor design relies heavily on fail-safe systems to try to ensure that there is no possibility of this happening in the event of a component or operational failure.

BREEDER REACTIONS

Uranium-235 most easily undergoes fission with slow (sometimes called thermal) neutrons. However, fast neutrons such as those it produces during fission can cause a reaction with uranium-238. In this case the result is not fission. Instead the uranium-238 atom captures the neutron to form uranium-239. This is unstable and rapidly decays, losing a beta particle (an electron) to form neptunium-239; neptunium is also unstable and loses a further beta particle to create plutonium-239.

$$^{238}_{92}U + n = {}^{239}_{92}U = {}^{239}_{93}Np + e^- = {}^{239}_{94}Pu + e^-$$

Plutonium is a fissile material like uranium-235 and will undergo a nuclear reaction when it absorbs a neutron. As a consequence of the reaction shown above, a certain amount of plutonium is generated in all nuclear reactors during their operation. A significant part of this undergoes fission, in the same way as uranium-235, releasing energy for power production. However, some remains in the spent fuel. This can be isolated if the fuel is reprocessed.

Since plutonium can be produced from uranium-238, and plutonium is a fissile material, it is possible to build a reactor that uses plutonium as its fuel, so long as enough of the fuel can be derived from uranium. If the reactor core of this plutonium reactor also contains uranium-238, then depending on the number of neutrons being produced in the core, it is possible to generate plutonium at the same time as producing power. This is the principle of the breeder reactor. In operating examples of this type of reactor uranium-238 is contained within a blanket that surrounds the core, where it can be irradiated with fast neutrons escaping the core. With careful design this type of reactor can produce more plutonium in the blanket surrounding the reactor that is used in the core—hence the name breeder.

Breeder reactors (sometimes called nuclear fast reactors because they usually exploit fast neutrons) use a coolant that is not a very efficient moderator and therefore does not slow down the neutrons significantly. This is typically liquid sodium although other moderators are possible. This coolant carries away the heat from the reactor but does not slow the neutrons in the core. These fast neutrons can still generate enough fission reactions in plutonium to form a sustainable nuclear reaction. Meanwhile the core of the reactor is surrounded with a blanket of uranium-238 and neutrons escaping the core react with uranium in this blanket to produce more plutonium. Eventually the uranium in the blanket is processed to isolate the plutonium, which can be used as further fuel for the reactor, and to produce yet more plutonium.

Another type of breeder reactor is based on the use of thorium. Thorium-232, the naturally occurring form, is of no use as a nuclear fuel because it is not fissile. However, if it is exposed to neutrons it reacts to form proactinium-233 which then decays in a second reaction to produce uranium-233.

$$^{232}_{90}\text{Th} + \text{n} = {}^{233}_{90}\text{Th} = {}^{233}_{91}\text{Pa} + \text{e}^- = {}^{233}_{92}\text{U} + \text{e}^-$$

Uranium-233 is a good fissile material, like uranium-235. One advantage of thorium over uranium-238 for breeder reactors is that it will react with slow neutrons and so can be introduced into conventional reactors in order to generate uranium-233.

There are, therefore, three fissile isotopes that can be used in nuclear reactors: uranium-233, uranium-235, and plutonium-239.

FUSION

The alternative energy-yielding nuclear reaction to fission is fusion. Fusion is the process that generates energy in the Sun and stars. In the Sun, hydrogen atoms combine to produce deuterium (heavy hydrogen) atoms and then deuterium and hydrogen atoms fuse to produce helium with the release energy. The reaction takes place at 10–15 million degrees Celsius and at enormous pressure.

The conditions in the Sun cannot be easily recreated on Earth, although fusion of the type taking place within the Sun has been achieved in laboratories. However for the purposes of electricity generation, another fusion reaction offers more potential because it takes place under more benign conditions than those in the Sun. This is the reaction between two isotopes of hydrogen, deuterium, and tritium, shown schematically in Fig. 3.2. Deuterium ^2_1H is found naturally in small quantities in water while tritium ^3_1H is made from lithium. These two will react to produce helium and energy.

$$^2_1H + ^3_1H = ^4_2He + n + @18 \text{ mev}$$

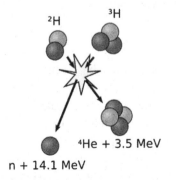

^2H ^3H

$^4He + 3.5 \text{ MeV}$

n + 14.1 MeV

Figure 3.2 Nuclear fusion reaction between deuterium and tritium. Source: Wikipedia commons.

The reaction between deuterium and tritium will only take place at 100 million degrees Celsius (but at much lower pressure than in the Sun). At this temperature all the atoms separate into a sea of nuclei and electrons, a state called a plasma. Since the constituents of a plasma are all charged, either positively or negatively, they can be controlled and contained using a magnetic field. This is crucial since there is no material that can withstand temperatures this severe. The most promising magnetic field for containing a plasma is toroidal and this has formed the basis for most fusion research. There is an alternative method of containing a fusion plasma called inertial confinement. This relies on generating extreme conditions within a small charge of tritium and deuterium, in essence creating a tiny sun in which the fusion takes place too fast for the particles to escape. Both systems of confinement are being developed for power generation.

NUCLEAR POWER PLANT CONVENTIONAL ISLAND COMPONENTS

The reactor in a nuclear power station forms the heart of the nuclear island of the plant. However the plant cannot produce electric power without its conventional island too. The conventional island of a nuclear power plant encompasses the components that make up the steam cycle of the plant. These include the steam generator, the steam turbines, the condenser, and system pumps. There will also be a range of other ancillary equipment including the control systems and the transformer substation that steps-up the voltage from the plant generators to that required by the grid.

The two most important components in the steam cycle are the steam generator and the steam turbines. Depending on the power plant design, the steam generator may be within the reactor vessel—as in a boiling water reactor—or it may be external as in a pressurized water reactor. In either case the thermal conditions in commercial nuclear power plants are such that the steam will be produced at relatively low pressure and temperature compared to that in most fossil fuel plants. Typically the steam pressure will be less than 80 atmospheres and around 300°C. This limits the overall thermodynamic efficiency that can be achieved in the steam turbine to around 30−33%.

Commercial nuclear power plants are usually large and conse-quently produce a very large amount of energy in the form of hot steam at these mild steam conditions. Typical operating plants have capacities of around 1000 MW and modern advanced designs can be up to 1500 MW or more. In order to extract the maximum amount of energy under these mild steam conditions requires that the turbines must be able to accommodate a large steam flow and this in turn means that steam turbines must be extremely large. In addition, the relatively low steam temperature means that the steam from the steam generator in a nuclear power plant is "wet"; wet steam contains small droplets of water.

Wet steam is a problem for steam turbines as it leads to water drop erosion of high-speed steam turbine blades. The higher the turbine blade tip speed, the greater the erosion. In order to minimize this, most nuclear power plants use half-speed turbine generators, so that for a 50-Hz system the steam turbine generator operates at 1500 rev/m instead of 3000 rev/m. The lower rotational speed reduces the turbine blade tip speed. The generator in such a plant must then be a four pole machine to achieve the necessary frequency.

In order to accommodate the large steam flow, there will usually be one high-pressure or high-pressure/intermediate-pressure turbine that takes the flow directly from the steam generator. The steam exiting this turbine will then be dried before passing into four or more low-pressure steam turbines. These will be extremely large turbines, often with last stage turbine blades of up to 1 m in length. All the turbines will be mounted on a single shaft driving the 1500-rev/m generator. The steam exiting the low-pressure turbines is then con-densed, usually using water, before returning to the steam generator.

Nuclear power plant efficiency could be increased if the plants could provide hotter, higher pressure steam. Research into nuclear configurations that can generate supercritical steam[2] at conditions similar to those found in modern coal-fired power stations is under-way. This could increase steam cycle efficiency in a nuclear power station to 45% or more.

[2]Supercritical steam is steam at such a high temperature and pressure that the difference between the gaseous and liquid forms ceases to exist.

CHAPTER 4

Water-Cooled Reactors

Nuclear reactor is the name given to the device or structure in which a controlled nuclear reaction takes place. There are a number of different reactor designs in use, but all have many features in common.

The core of the reactor is its heart, the place where the nuclear fuel is placed and where the nuclear reaction takes place. The fuel is most frequently formed into pellets roughly 2 cm in diameter and 1–2 cm long. These pellets are loaded into a fuel rod, a hollow tube of a special corrosion-resistant metal; this is frequently a zirconium alloy which is transparent to neutrons. Each fuel rod is 3–4 m long and will contain 150–200 pellets. A single reactor core may contain up to 75,000 such rods. Fuel rods must be replaced once the fissile uranium-235 they contain has been used up.[1] Refueling a reactor is a lengthy process which can take as much as three weeks to complete during which the reactor normally has to be shut down. However, some designs allow refueling while in operation.

Inside the core, in between the fuel rods, there are control rods, made of boron, boron steel or boron carbide, which are used to control the nuclear reaction. Boron is a good absorber of neutrons. These rods can be moved in and out of the core and they will be of different types. Some may be designed to completely stop the reaction in the core, others to adjust the speed of the reaction. The core will also contain a moderator to slow the neutrons released by the fission of uranium atoms. In many cases the moderator is also the coolant used to carry heat away from the core.

The outside of the core may be surrounded by a material that acts as a reflector to return some of the neutrons escaping from the core. This helps maintain a uniform power density within the core and allows smaller cores to be built. There may also be a similar reflecting material in the center of the core.

[1]Usually there is still some uranium-235 left but too little to sustain the reaction in the core.

The coolant collects heat within the core. This coolant may be the thermodynamic fluid that drives the plant turbine—as in a boiling water reactor (BWR)—or it may transfer its heat through an external heat exchanger to a secondary circuit where it is exploited to raise steam to drive a steam turbine—as in a pressurized water reactor (PWR). The coolant may be water (light water), deuterium (heavy water), a gas such as helium or carbon dioxide or a metal such as sodium. As noted in Chapter 3 the core and its ancillary equipment are normally called the 'nuclear island' of a nuclear power plant while any external steam generator,[2] the steam turbine and generator are called the 'conventional island'. The coolant/steam cycle will link the nuclear and conventional islands.

A nuclear power plant will also contain a host of components to ensure that the plant remains safe and can never release radioactive material into the environment. The most important of these is the containment. This is a heavy concrete and steel jacket which completely surrounds the nuclear reactor. In the event of a core failure the containment should be able to completely isolate the core from the surroundings and remained sealed, whatever happens within the core. It should also be able to resist a massive external impact such as an aircraft crashing into the structure.

WATER-COOLED DESIGNS

Although there have been a range of reactor designs used commercially since the nuclear power age began, by far the majority of those built use water as both the moderator and the coolant. Water is a good moderator for slowing fast neutrons because the hydrogen atoms in water are the same size as a neutron. If the moderator was composed of very large atoms, then the neutrons would simply bounce off them; if the moderator atoms were much smaller than the neutron, then the latter would simply push them aside without transferring significant kinetic energy during a collision. When the masses are similar, then the optimum conditions for momentum and energy transfer are achieved.

When a fast neutron collides with a proton, it can transfer a large amount of energy to its collision partner, reducing its own kinetic energy in the process. This process in effect transfers excess energy

[2]In some designs the primary steam generator is within the core.

from the neutrons to the water as heat. The heat can then be carried away by the water through the cooling circuit that forms part of the steam generation system of the power plant.

There are two primary types of water-cooled reactor, a BWR and a PWR. The latter includes two types, a pressurized light water reactor and a pressurized heavy water reactor. In addition, there is a Russian-designed reactor that uses graphite as its moderator but water as the coolant.

THE BOILING WATER REACTOR

The BWR is one of the two most important reactor designs in use today. The first BWR was an experimental reactor called Borax I which was built at the Argonne National Laboratory in Idaho, United States in 1952. The purpose of the experiment was to determine if allowing coolant water to boil within the core of the reactor would lead to core instability. The experiment was successful—there was no instability when the water boiled—and the initial unit was upgraded to Borax II and then, in 1954, Borax III was built. The latter is famous for having supplied the first nuclear electric power in the United States, to the city of Arco. In fact the unit was not originally designed to provide power but was rapidly modified in early 1955 to counter a Russian announcement of a 5-MW nuclear power plant that has been built in the USSR. Borax III had a generating capacity of about 2 MW.

The technology developed for the Borax reactors was subsequently adopted by General Electric Co in the United States which developed a commercial nuclear reactor based on the boiling water concept. The first of these reactors, the Dresden Nuclear Power Plant in Illinois, USA, with a generating capacity of 210 MW, began operating in 1960 and is considered the first commercial nuclear power plant in the United States. There were 78 BWR reactors in operation around the world at the end of 2015.

The BWR uses ordinary water (light water) as both its coolant and its moderator. Its unique feature is that steam is generated directly inside the reactor core. In the BWR, the water in the reactor core is permitted to boil under a pressure of 75 atm, raising the boiling point to 285°C and the steam generated is taken from the core and used directly to drive a steam turbine. This steam is then condensed and

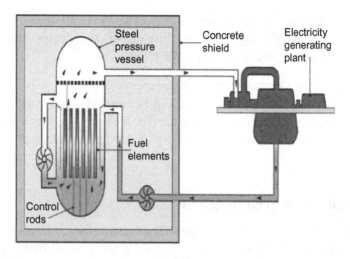

Figure 4.1 Boiling water reactor. Source: The Institution of Engineering and Technology Nuclear Factsheet.

recycled back to the reactor core. This is shown in Fig. 4.1. Since the steam is exposed to the core, there is some radioactive contamination of the turbines but this is short-lived and turbines can normally be accessed soon after shutdown.

The BWR configuration represents probably the simplest possible for a nuclear reactor because no additional steam generators are required. However the internal systems within a BWR are complex. The BWR uses enriched uranium as its fuel, with an enrichment level of around 2.4% uranium-235. This fuel is placed into the reactor in the form of uranium oxide pellets in zirconium alloy tubes. There may be as much as 140 tonnes of fuel in 75,000 fuel rods. Refueling a BWR involves removing the top of the reactor. The core itself is kept under water, the water shielding operators from radioactivity. Boron control rods enter the core from beneath the reactor. In modern BWRs the control rods are used to keep power generation within the reactor core homogeneous and to compensate for consumption (burn-up) of the fuel. The rate of water flow through the core is then used to control power. However some early BWRs used only natural water circulation with no pumps. In these reactors the control rods had to be able to control the power between 0% and 100%.

There is also a natural moderation feedback process that takes place within the BWR core. The steam above the water in the core is less good at moderating neutrons, so if the core overheats and generates more steam, displacing some of the water around the core, this will tend to reduce the rate of fission.

In common with all reactors the fuel rods removed from a BWR reactor core are extremely radioactive and continue to produce energy for some years. They are normally kept in a carefully controlled storage pool at the plant before, in principle at least, being shipped for either reprocessing or final storage.

Most BWR reactors are typically 900–1100 MW in generating capacity, with an efficiency of 32%. Early plants, mostly entering service in the early 1970s, were around 500–600 MW in capacity. Advanced BWR designs have capacities of up to 1400 MW and an efficiency of around 33%. A small number of these are in operation in Asia.

THE PRESSURIZED WATER REACTOR

The PWR is the second of the two key nuclear power technologies in use today. It emerged from research into submarine propulsion units that took place during and after the Second World War. A prototype was built at the Idaho National Laboratory in the United States in 1953 and the first US submarine with a nuclear power unit, USS Nautilus, was launched in 1955. The technology was taken up by Westinghouse, which went on to build many of the US Navy's submarine propulsion units, and was developed for power generation. A second US company called Combustion Engineering also became involved in PWR development and was a rival to Westinghouse. Its design eventually became merged with that of Westinghouse.

The first PWR power plant in the United States was the Shippingpoint Atomic Power station. However, this was based on a canceled nuclear aircraft carrier power unit and was of extremely unusual design with highly enriched uranium (93% uranium-235) seed fuel surrounded by a blanket of natural uranium-238. The unit went critical in 1957 and had a power generating capacity of 60 MW. As a result of its unusual design, Shippingport is often considered a demonstration PWR rather than the first commercial nuclear power plant in the United States. That honor instead belongs to Dresden nuclear power plant (see above), a boiling water design which started in 1960. The first commercial PWR was the Yankee Rowe nuclear power station which also started in 1960, but slightly later than Dresden. This plant had a generating capacity of

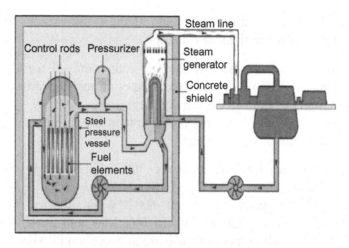

Figure 4.2 Pressurized water reactor. Source: The Institution of Engineering and Technology Nuclear Factsheet.

185 MW. At the end of 2015, there were 283 PWRs operating around the world, making these the most common reactors in use.

The PWR also uses light water as both coolant and moderator. However, in the pressurized water system the cooling water is kept under pressure so that it cannot boil. As a consequence the PWR differs in another respect from the BWR; the primary coolant that flows through the reactor core does not drive the steam turbine. Instead heat from the primary water cooling system is captured in a heat exchanger and transferred to water in a secondary system, as shown in Fig. 4.2. It is the water in this second system which is allowed to boil and generate steam to drive the turbine.

The core of a PWR is filled with water, pressurized to 150 atm, allowing the water to reach 325°C without boiling. Containing this high pressure over the volume of the core presents, one of the major engineering challenges in PWR design. The use of a second water cycle introduces energy losses which make the PWR less efficient at converting the energy from the nuclear reaction into electricity. On the other hand the higher steam temperature and pressure allow for greater thermodynamic efficiency and typical overall efficiency, at 32%, is similar to a BWR. In addition the primary/secondary cooling circuit arrangement has other advantages regarding fuel utilization and power density, making it competitive with the BWR. High-pressure core operation with no steam production allows the reactor to be more compact.

The PWR uses uranium fuel enriched to 3.2%, a slightly higher enrichment level than in a BWR. This is responsible for a higher power density within the reactor core. As with the BWR the fuel is introduced into the core in the form of uranium oxide pellets. A typical PWR will contain 100 tonnes of uranium in around 50,000 fuel rods and 18 million fuel pellets. Refueling is carried out by removing the top of the core. However, in a PWR the control rods are inserted from above too, allowing gravity to act as a fail-safe in the event of an accident.

A typical PWR has a generating capacity of 1000 MW although early plants built in the 1970 s were much smaller. The original Westinghouse design is the starting point for most western PWRs. France also developed a PWR which was originally based on the Westinghouse design, but the designs later diverged so that the French PWR is now an independent design. Advanced PWR designs, some of which are being built today, range in size from 1100 to 1700 MW.

THE PRESSURIZED HEAVY WATER REACTOR

While most water-cooled reactors use light water (H_2O) as their coolant, one design, the Canadian deuterium uranium (CANDU) reactor, uses heavy water. The CANDU reactor was developed in Canada with the strategic aim of enabling nuclear power to be exploited without the need for imported enriched uranium. Canada is home to some of the richest uranium ores in the world but uranium enrichment is an expensive and highly technical process. If it can be avoided, countries such as Canada with natural uranium reserves can more easily exploit their indigenous reserves to generate energy. This has made the CANDU reactor, which uses unenriched uranium, attractive outside Canada too.

The CANDU reactor uses, as its moderator and coolant, a type of water called heavy water. Heavy water is a form of water in which the two normal hydrogen atoms have been replaced with two of the isotopic form, deuterium. Each deuterium atom weighs twice as much as a normal hydrogen atom, hence the name heavy water. Heavy water occurs in small quantities in natural water.[3]

[3] A deuterium atom is a hydrogen atom with an extra neutron, giving it twice the mass of normal hydrogen. About one in every 6760 naturally-occurring hydrogen atoms is a deuterium atom. The two can be separated using electrolysis which selectively splits normal water but leaves heavy water.

Heavy water has to be separated from natural water, so it is much more expensive than light water but it has the advantage that it absorbs fewer neutrons than light water. As a consequence, it is possible to sustain a nuclear reaction using heavy water cooling without the need to enrich the uranium fuel. The CANDU reactor has the additional advantage that it can be refueled without the need to shut it down; in fact this is necessary with natural uranium fueled reactor to keep the plant going. Avoiding lengthy refueling shutdowns provides better operational performance.

The CANDU fuel is loaded in the form of uranium oxide pellets housed in zirconium alloy rods that are inserted horizontally into pressure tubes penetrating the core instead of vertically as in other PWRs and BWRs. Fuel replacement involves pushing a new rod into a pressure tube which passes through the vessel containing the heavy water (called a calandria) and forcing the old tube out of the other end. The pressure tube must be isolated from the heavy water so that refueling can be carried out without the need to shut down the reactor.

The heavy water coolant in the CANDU reactor is maintained under a pressure of around 100 atm, lower than is a light water PWR, and the water reaches around 290°C without boiling. Heat is transferred through a heat exchanger to a light water system with a steam generator and the secondary system drives a steam turbine in much the same way as a PWR (see Fig. 4.3). Efficiency is slightly lower than light water reactors at around 30%.

The CANDU reactor was developed by Atomic Energy of Canada and that country has the largest CANDU fleet but reactors have also been supplied to countries such as Argentina, South Korea, India and Pakistan. Reactor capacity is typically 600–700 MW, smaller than the alternative light water reactors. There are 49 in operation.

THE VVER REACTOR

The Russian VVER reactor (the abbreviation comes from Vodo-Vodyanoi Energetichesky Reaktor which translates as Water-Water Power Reactor, i.e., a water-cooled, water-moderated reactor) is a Russian-designed variant of the PWR. The original version of this design to enter service was the VVER-440 with a generating capacity

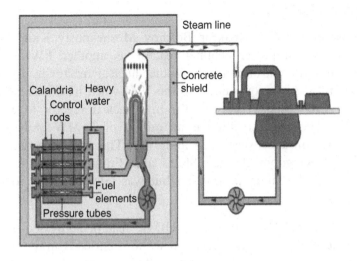

Figure 4.3 Pressurized heavy water reactor. Source: The Institution of Engineering and Technology Nuclear Factsheet.

of 440 MW. The first reactors of this type were built at Novovoronezh in 1972 and 1973. Soon after this, boron control rods were introduced, allowing the number of control rods to be reduced and leading to a modified design called V-230. A third design, the V-213 was introduced later in an attempt to apply modern safety standards to VVER nuclear units. Alongside the VVER-440, Russia also developed the VVER-1000, a 1000-MW version based on the same design concept. The first of these entered service at Novovoronezh in 1980.

The VVER typically operates with a coolant pressure of 150 atm and a temperature of 300°C, similar to a western PWR. However, there are a number of minor design variations compared to the typical Westinghouse design. Fuel assemblies in the Russian design are hexagonal rather than square, in principle allowing closer packing of the rods, and the steam generators are oriented horizontally rather than vertically. However, like the Westinghouse PWR, control rods are inserted from above the reactor vessel. The uranium fuel for the VVER reactor is enriched to between 2.4% and 4.4% uranium-235.

There are 23 VVER-440 reactors in operation and 28 VVER-1000 units. A VVER-1200 with an output of 1200 MW is the latest evolution of this design and units to this specification are being built in Russia as well as being offered for construction in the rest of the world.

THE RBMK REACTOR

The Russian RBMK (Reaktor Bolshoy Moshchnosty Kanalny that translates as high-power channel reactor) is another BWR, but of significantly different design to any other nuclear reactor in use today. The design of this reactor dates to the 1954 when its predecessor, the Obninsk AM-1 reactor, became the first nuclear reactor to generate electricity. The original unit has an output of 30 MW and supplied power to Obninsk between 1954 and 1959. This unit, like reactors in the United Kingdom, United States and France at the time, was intended for plutonium production but was later modified to power generation use.

Like the CANDU reactor, the RBMK reactor addressed the need create a sustainable nuclear reaction without enriched uranium, or in this case with low uranium enrichment. This requires a moderator that has a low absorption coefficient for neutrons. In many early and experimental nuclear reactors, graphite was used because it has a low absorption coefficient and also because it was cheap. Deuterium, while it absorbs fewer neutrons, is much more expensive to produce.

The RBMK reactor was essentially a utilitarian design. the core is loaded with graphite blocks through which holes were cut to allow the fuel rods and control rods to be inserted. Also running through the core are pressure tubes that carry water pumped from a reservoir below the reactor. This is allowed to boil to steam within the pipes and is then collected at the top and fed to steam generator vessels. The water/steam system is pressurized to 68 atm and the water reaches a temperature of 284°C. During operation the graphite moderator reaches a temperature of 730°C. The vessel containing the core is filled with a helium/nitrogen gas mixture which is inert, to prevent oxidation of the graphite and also helps heat transfer between the graphite and the water pressure tubes. A cross-section of an RBMK reactor is shown in Fig. 4.4.

The RBMK reactor has no containment vessel in the sense understood by modern reactor designers. Instead it is placed inside a reinforced, concrete lined cavity. The reactor core sits on a steel plate and is surrounded by a steel casing. The conventional island for this plant which includes the steam separators is located in a concrete housing.

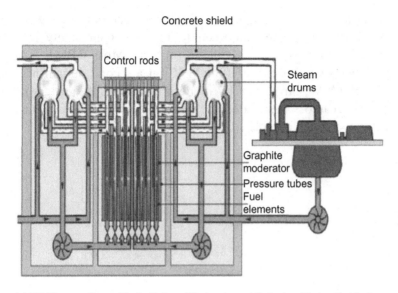

Figure 4.4 RBMK reactor. Source: The Institution of Engineering and Technology Nuclear Factsheet.

Uranium for use in the RBMK reactor is enriched to 1.8–2.0%. Control rods in the core consist of two types, both based on boron. A number of short control rods are inserted from below to help create a uniform power distribution. The main control rods for controlling the reaction are inserted from above.

The RBMK reactor has a flaw in its design that eventually led to a catastrophic failure. In a traditional BWR, if the core overheats more water is turned into steam. This reduces the amount of water around the core, reduces the moderating of neutrons and so reduces the number of fission reactions and the amount of power production. The RBMK has a graphite moderator but a boiling water coolant. If the core overheats, more steam is again produced. In this case, however, the steam does not significantly affect the degree of neutron moderation which is still controlled by the graphite but it does reduce the absorption of neutrons by water since steam absorbs fewer neutrons that light water. This leaves more neutrons to react in the core, leading to more heat and a potential runaway situation. It was a situation of this type that led to the catastrophic failure of one of the RBMK reactors at the Chernobyl nuclear power plant in 1986.

Since that accident a number of RBMK reactors have been shut down. Others have been modified to improve their safety features. Eleven of these reactors are still operational. Most RBMK reactors had a generating capacity of 1000 MW but a second design increased this to 1500 MW, large even by modern standards. A new generation of RBMK reactors designated the MKER reactor was designed but none has even been completed.

CHAPTER 5

Gas-Cooled Reactors

The gas-cooled reactor forms a second branch of early nuclear power reactor design. Like the heavy water PWR the aim of this reactor design was to be able to utilize natural, unenriched uranium as fuel. After the Second World War, uranium enrichment technology in the west was virtually all in the hands of the United States. As a consequence other countries looked at ways of developing nuclear reactors that could operate without enriched uranium. The result, in France and in the United Kingdom, was generations of gas-cooled reactors with graphite moderators. These all used carbon dioxide as the coolant gas.

Development in the United Kingdom led to a first generation of gas-cooled reactors called Magnox reactors. These were later superseded by a new design called the advanced gas-cooled reactor (AGR). In France a series of independently designed reactors of a similar type were designed. As in the United Kingdom a second generation was explored but France decided instead to opt for PWR technology.

There were also experiments in the United States with a gas-cooled reactor called the ultrahigh-temperature reactor experiment (UHTREX). This experimental reactor operated from 1959 until 1971 and was part of a program to see if there were advantages to using unclad nuclear fuel instead of fuel contained inside fuel rods. The core of this reactor was constructed of graphite which also acted as the moderator while the coolant was helium. The reactor had a rating of $3\,MW_{th}$ and operated up to $1316°C$, a much higher temperature that any commercial reactor. However the design was finally abandoned and development in the United States was stopped.

Russia also experimented with a gas-cooled reactor called the KS 150. A single 143 MW unit was constructed in Czechoslovakia and commissioned in 1972. The reactor was novel in using heavy water as the moderator and carbon dioxide as the coolant. The unit suffered a series of accidents and was decommissioned in 1979.

THE MAGNOX REACTOR

The Magnox reactor was a nuclear technology developed in the United Kingdom for both weapons manufacture and power production. The first reactor of this type was built at Calder Hall in Cumbria. It was a dual purpose reactor with a generating capacity of 50 MW and is now considered the first commercial nuclear power station in the world although the power was essentially a by-product of the plutonium production process.

The reactor used carbon dioxide as its coolant and graphite as the moderator. Fuel was natural, unenriched uranium metal which was loaded into magnesium-aluminum alloy fuel rods from which the name Magnox is derived. The control rods were made from boron steel. The carbon dioxide was pressurized to around 20 atm and the hot gas exited the core at a temperature of 360°C. A diagram of a Magnox reactor is shown in Fig. 5.1. Overall efficiency of the Magnox reactor was relatively low at around 18% although the steam cycle efficiency was higher at 31%. The use of unenriched uranium meant that the cores had to be refueled while online in order to maintain the nuclear reaction.

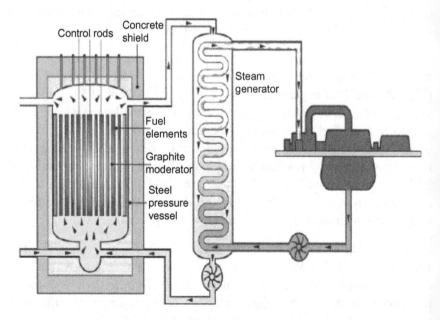

Figure 5.1 Magnox reactor. Source: The Institution of Engineering and Technology Nuclear Factsheet.

Although a series of Magnox reactors were built, the design was never standardized and each new plant had slight modifications compared to those that had gone previously. This meant that each unit was essentially a prototype, making them costly to build since no standard design could be rolled out. Early units had steel pressure vessels, while some later ones had pre-stressed concrete vessels with steel liners. Shapes varied too; some of the vessels were spherical, others were cylindrical. The earliest plants were all 50 MW in capacity but size later increased so that the final Magnox plant, which began operating at Wylfa in Wales in 1971, had units with a capacity of 490 MW. In total 26 Magnox reactors were built at 11 power plants in the United Kingdom. One was exported to Japan, and another to Italy. Ironically, North Korea also developed a Magnox reactor based on the UK design after the latter had been made public at an *Atoms for Peace conference.*

The layout of the early Magnox reactors kept the reactor alone inside the pressure vessel while the primary coolant circuit ran through a heat exchanger and steam generator outside the vessel. In later designs the heat exchanger and steam generator were inside the pressure vessel containing the core. Here the hot carbon dioxide was used to heat water and raise steam to drive the plant steam turbine. There was no containment as found in modern nuclear power plants because the design was at the time considered inherently safe. This proved to be an overconfident assessment when there was a fuel melt-down in a channel of the core at a plant in 1967, an accident which led to the release of radiation. In addition the steel vessels in the earlier plants of this design let significant amounts of neutron and gamma radiation pass, creating a hazard for local people.

One of the major limiting factors with the Magnox plants was the magnesium–aluminum alloy. This alloy was used because it had a low level of neutron capture but it could not operate at a very high temperature, limiting the overall thermodynamic efficiency that could be achieved. In addition the cladding reacted with water so that long-term storage in fuel-cooling ponds was not possible and the fuel had to be reprocessed. The last Magnox reactor in the United Kingdom closed in 2015.

THE UNGG REACTOR

The French réacteur nucléaire à l'Uranium Naturel Graphite Gaz (UNGG)—nuclear reactor with natural uranium, graphite and gas—was a French reactor design that was developed in parallel to the UK's Magnox reactor programme. As with the latter, the nuclear design used unenriched uranium as fuel, with a graphite moderator and carbon dioxide gas coolant. And as in the United Kingdom, these plants were able to produce plutonium for weapons as well as power.

In the initial UNGG design the nuclear fuel was distributed within the graphite core in the form of slugs of a uranium alloy. Later a design using fuel rods was developed. Fuel rods were loaded horizontally into the core at first, but vertical fuel rods were introduced for later designs. The fuel rod cladding was made from a zirconium—magnesium alloy which had the same flaw as the UK magnesium—aluminum alloy, the material reacted with water.

The cores of the first plants were enclosed in steel vessels. However a concrete containment, several meters thick, was later adopted. In some plants the carbon dioxide circuit and gas/water heat exchanger was within this containment, in others only the core was inside and the heat exchanger was external.

Nine reactors of this type were built in France between 1950 and 1960. The first three were built by the French Atomic Energy Commission. Following that, six further reactors were built by Eléctricité de France (EDF). One UNGG was also built in Spain and another in Israel. As with the UK programme, the design evolved from unit to unit. The largest of them had a generating capacity of 540 MW.

The construction of the French reactor fleet was part of a strategy of the French government at the time, led by General de Gaulle, to become energy independent. However, in 1969, EDF proposed adopting an American PWR design for future nuclear reactors and this became official French policy at the end of that year. The first three of these reactors were shut down between 1968 and 1987. The other six have also been withdrawn from service and are being decommissioned.

The French experience highlighted a number of limitations of this particular design. Carbon dioxide could cause steel corrosion at high temperatures. Meanwhile the graphite core moderator also presented problems depending on the temperature at which it was operated. In this case a higher temperature was better. Finally the steam generators in these plants had shorter lifetimes than the reactors themselves and so needed to be replaced during the plant lifetime.

THE ADVANCED GAS-COOLED REACTOR

The AGR was a second generation of graphite moderated, carbon dioxide-cooled reactor developed in the United Kingdom. The Magnox reactors had not proved very efficient and in order to improve efficiency, a new design was developed which operated at higher temperatures and pressures and with a higher power density within the core in order to reduce capital costs and improve the economics. To achieve this the Magnox fuel rod cladding was changed to stainless steel and the uranium metal fuel was changed to uranium dioxide. The fuel change then required that the uranium be enriched and an enrichment level of 2.3% was used.

The design broadly followed that of the former Magnox reactor with a graphite moderator. As with the later reactors of the Magnox design, the steam generator was placed within the concrete containment of the reactor. Fuel rods and control rods were inserted from above as shown in Fig. 5.2. The design also allowed for refueling to be carried out while the reactor was in service.

The operating pressure of the coolant gas was 45 atm and the temperature of the carbon dioxide entering the hottest section of the steam generator was 619°C, much hotter than in the Magnox reactor. This allowed steam to be produced at a pressure of 165 atm and a

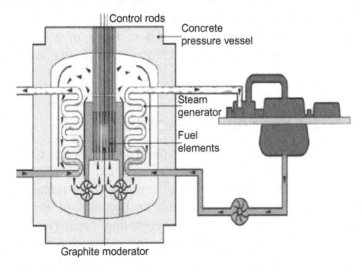

Figure 5.2 Advanced gas-cooled reactor. Source: The Institution of Engineering and Technology Nuclear Factsheet.

temperature of 538°C at the inlet to the high-pressure steam generator. The hotter, higher pressure steam meant that the steam cycle could operate at 42% efficiency and this was one of the principal design aims of the AGR. These steam conditions were chosen to be similar to conventional coal-fired power plants in use at the time so that the same steam turbine generators could be used in both.

Uniquely among nuclear power plants, the AGR design included two reactors providing steam to a single set of steam turbines. Each reactor had a core with 332 fuel rod channels and 89 control rod channels. The core contained around 117 tonnes of uranium. The gross power output of each reactor was 660 MW for a total generating capacity for each plant of 1320 MW.

The UK government hoped that the AGR design would prove a competitor for the US PWR and BWR reactors. Five twin reactor power plants were ordered in quick succession but the design proved complex. The first plant at Dungeness was ordered in 1965 but did not enter service until 1983, 18 years later. The design was modified for subsequent plants and these entered service earlier than the first. Two further plants were ordered, so that a total of 7 plants, and 14 reactors, eventually entered service. All continue to operate.

In spite of their relative success the plant turned out to be costly and no orders for plants outside the United Kingdom were taken. When the United Kingdom decided to built its next nuclear power plant, the AGR design was abandoned in favor of a US-designed PWR. This was commissioned in 1995 after a 7-year construction programme.

HEAVY WATER GAS-COOLED REACTORS

In 1962 the Atomic Energy Commission of France began construction of an experimental heavy water moderated gas-cooled reactor (HWGCR). The plant was seen as a prototype for a replacement for the UNGG. The prototype was built at Brennilis in Brittany and entered service in 1967. It had a generating capacity of 70 MW and used water from the local river Ellez for condenser cooling.

The replacement of graphite moderation with heavy water was intended to overcome the limitations of the graphite while allowing a gas-cooled reactor to operate with unenriched uranium. However the French government's decision in 1969 to adopt the US PWR design as the basis for future French reactors ended research into a new generation of gas-cooled reactors in France. The plant continued to operate until 1985 when it was finally closed and it is now in the process of being decommissioned.

Russia also experimented with an HWGCR, a design called KS 150. Work on the design began in the 1950s and construction, in Czechoslovakia, commenced in 1958. However the construction took 16 years and the plant only entered service in 1972. The prototype power station used uranium metal as fuel, and this was loaded into fuel rods made of a magnesium beryllium alloy. The core was a carbon steel cylinder that contained the heavy water, with channels in it for the fuel rods to pass through. Carbon dioxide could circulate through these fuel rods to carry away the heat generated. Operating pressure was around 54 atm and the carbon dioxide coolant reached 426°C. The overall efficiency of the plant was estimated to be 18.5% and it had a generating capacity of 143 MW.

The plant suffered a major accident during refuelling in 1977. This was the culmination of a series of problems and the plant was decommissioned in 1979 and a second unit of the same type canceled.

CHAPTER 6

Breeder Reactors

Breeder reactors are a special type of nuclear reactor that can produce additional nuclear fuel at the same time as producing energy for electricity generation. All energy producing nuclear reactors rely on one of a small number of fissile elements to generate energy. These elements, discussed in Chapter 3, have nuclei that will react with neutrons and split into two smaller nuclei with the release of large quantities of energy. The only fissile nucleus that occurs naturally in significant quantities is uranium-235 which makes up roughly 0.7% of the uranium found in uranium ores.

There are two other fissile isotopes of value for energy production, plutonium-239 and uranium-233, but neither of these can be found naturally. However, both of them can be produced in nuclear reactions of a different type to those that result in fission. The most important of these alternative processes is a reaction between uranium-238 and a fast neutron that produces plutonium-239. Since uranium-238 is much, much more abundant on the Earth than uranium-235, this reaction offers a way of providing a much larger source of fissile material that the naturally occurring fissile uranium-235 isotope can provide alone.

The second breeder reaction involves thorium-232 that will react with a neutron to produce uranium-233. Thorium-232 is even more abundant than uranium-238 and so could in principle provide an even large source of fissile material. However, this reaction has yet to be exploited on a large scale for nuclear energy production. In contrast the production of plutonium has been widely explored because it is also a key element in nuclear weapons. The two isotopes, uranium-238 and thorium-232, which will react with neutrons to produce new fissile isotopes are called fertile isotopes.

The development of plutonium breeder reactors can be traced to the earliest days of the nuclear age and like other reactors these were a product of weapons development. Furthermore the potential to build reactors that could produce more fuel than they consumed was an attractive one during the early years of nuclear power when the availability of uranium was uncertain. During the two decades following World War II, the United States, the United Kingdom, France, Germany, Japan, the Soviet Union, and India all set up breeder reactor research programs. By the middle of the 1980s, with the development of nuclear power slowing and the abundance of uranium much higher than had previously been thought, the need for breeder reactors became less urgent and for most countries this research was no longer a priority. However, both India and Russia have continued development as a result of domestic fears of a uranium shortage, while China is building prototype units to a Russian design.

In spite of the wealth of research the problems encountered during development of breeder reactors have meant that no commercial unit has ever entered service, although research reactors have generated electricity. However in the last two decades a renewed interest in fast neutron breeder reactors has been stimulated by the need to find a way of disposing of nuclear waste from existing power plants. Breeder reactors can burn the plutonium formed in conventional reactors, thereby eliminating at least part of the waste fuel disposal problem. They can also destroy the actinides[1] in radioactive waste from water-cooled reactors that cannot be reused for fuel. For this to be effective as a means of dealing with nuclear waste, there would need to be many more operating breeder reactors than exist today.

PLUTONIUM BREEDER REACTORS

The majority of breeder reactors, whether experimental, prototypes or demonstration plants, that have been built have been plutonium breeder reactors. These are also sometimes known as nuclear fast reactors or fast breeder reactors. Plutonium-239 is a fissile material and its nucleus will split when struck by a neutron, generally producing two nuclei of smaller elements and a number of fast neutrons. Plutonium-239 is the most common nuclear fuel used in fast breeder reactors and it provides both the source of energy for electricity production and a

---

[1]Actinides are the elements with atomic numbers from 89 (actinium) to 103. Isotopes of these high atomic number elements are often formed from uranium in nuclear fuel and many of them are highly radioactive.

source of fast neutrons. These fast neutron are then exploited both to generate further fission reactions and to react with uranium-238 which is also present in the reactor. Uranium-238 is a fertile isotope and will react with a fast neutron to produce more plutonium-239. In this way the breeder reactor can produce both energy and more fuel.

One of the key differences between a conventional nuclear reactor and a plutonium breeder reactor is that the latter does not have a moderator. In the conventional reactor the fast neutrons produced from uranium-235 fission reactions are slowed because slow neutrons are much more likely to react with further uranium-235 nuclei that are fast neutrons. Fast neutrons will react, but the probability of reaction is much lower. Plutonium-239 also reacts with both slow and fast neutrons but, critically, it has a higher probability of reaction with a fast neutron than uranium-235.

The fast breeder reactor requires a high density of fast neutrons because it is these that will react with uranium-238 and produce more plutonium. In order for a fast neutron reactor to achieve criticality, the core will contain a much higher percentage of fissile material—typically around 20% or more of plutonium-239—than would be found in a slow neutron reactor. This higher concentration allows a controlled chain reaction to be achieved with fast neutrons. Plutonium has a second advantage too, it produces around 25% more fast neutrons from each fission reaction than uranium-235 and this means there are more neutrons to share between fission and production of more plutonium.

The structure of a fast neutron reactor typically involves a core containing the enriched plutonium fuel, usually mixed with depleted uranium to achieve the required level of enrichment. The latter is the uranium-238 left from the enrichment of uranium and it is referred to as depleted because it has a much lower concentration of fissile uranium-235 than would be found in natural uranium. This uranium-238 within the core will produce some additional plutonium. However, in order to be able to make more plutonium that it burns, the reactor core is surrounded by a further blanket of depleted uranium. Stray fast neutrons from the core pass into this blanket and generate more plutonium. Moreover, the neutron density here is too low to lead to many fission reactions so most of the plutonium remains in the blanket, once produced.

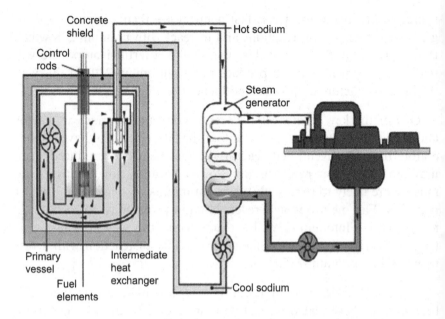

Figure 6.1 Sodium-cooled fast neutron reactor. Source: The Institution of Engineering and Technology Nuclear Factsheet.

The other key element of the fast neutron reactor is the coolant. Since the reactor uses fast neutrons the coolant cannot be either a moderator or a neutron absorber. The material that has proved the most popular coolant for fast neutron reactors is liquid sodium. A cross-section of a reactor of this type is shown in Fig. 6.1. Sodium has good heat-carrying properties and, importantly, does not absorb or slow neutrons. However the material is very reactive if exposed to air or water and so the cooling circuits have to be extremely strictly engineered. The core of a fast neutron reactor is usually smaller than that of a conventional slow neutron reactor and it has a higher power density within the core. This leads to a higher core temperature of 500–550°C. The core usually operates at atmospheric pressure, again unlike slow neutron reactors which usually operate at high pressure. But like the latter they have control rods to manage the nuclear reaction and these are made of boron carbide.

The liquid metal coolant in a fast neutron reactor is passed through a heat exchanger through which water is passed and steam generated. The steam is then used to drive a steam turbine for power production, in much the same way as a slow neutron reactor. As with the latter, the heat exchanger/steam generator may be located either inside the containment vessel that encloses the reactor core, or outside.[2] Fast neutron reactors require these same protective enclosures and the same safety features as conventional reactors.

While liquid sodium is the most popular coolant, others have been tested too. Candidates include liquid lead or a lead-bismuth mixture. Gas-cooled reactors, often using helium, are also possible. However the liquid sodium offers the best breeding potential. The latter is defined by the breeding ratio, a figure that shows how much new fissile material is produced for each unit of fissile material burnt. The figure must be greater than one if the reactor is to produce more fuel than it consumes. With a sodium-cooled reactor a breeding ratio of 1.3 can be achieved. With other coolants such as lead−bismuth, the breeding ratio is usually less than one.

The fuel that is loaded into a nuclear fast reactor is normally in the form of plutonium oxide and uranium oxide. These oxides do not react with sodium or lead but they have relatively low thermal conductivities. Alternatives with high thermal conductivity such a mixed metal fuel or fuels made from uranium and plutonium carbides of nitrides have also been tested but these present other problems that make them less easy to manage than the conventional oxide fuels.

In order to close the fuel cycle for a fast neutron reactor, the fuel and the blanket material from the reactor must be processed to isolate the plutonium so that it can be used to manufacture more fuel. While developing fast neutron reactors, several countries such as the United Kingdom and France also developed nuclear waste reprocessing facilities that are capable of carrying out the large-scale separation of plutonium for fuel manufacture. Today these processing plants are more likely to be used for waste fuel reprocessing from slow neutron reactors. The plutonium produced from the fuel is then used to make a mixed oxide fuel containing both fissile uranium and fissile plutonium. This can be used as fuel in some conventional reactors. However, it could be used in breeder reactors in the future.

[2]With a liquid sodium−cooled reactor, there may be two sodium loops and a sodium/sodium heat exchanger in addition to the sodium/water heat exchanger. In this type of design the first loop is usually contained within the reactor core.

FAST NEUTRON REACTOR PROJECTS, PAST AND PRESENT

According to the World Nuclear Association, around 20 fast neutron reactors have operated since the 1950s. Most of these have been experimental or pilot schemes but a small number have been demonstration projects that have produced electric power for delivery to the grid.

In the United States, the first experimental breeder reactor called EBR-1 began operating in Idaho in 1951. The plant used a sodium—potassium alloy as coolant and metallic uranium fuel. It was able to produce around 200 kW of power, enough to power the facility and ran until 1963. It was succeeded by EBR-II, a liquid sodium research reactor with a power output of 20 MW, which ran until 1994. This reactor also formed the basis for a project called the integral fast reactor which was intended to integrate a reactor, reprocessing and fuel manufacture in a single facility, again using enriched metallic uranium fuel. The only commercial breeder reactor in the United States was the Fermi 1 demonstration project in Michigan. The project started in 1963 but the plant only operated for 3 years before coolant problems forced it to close. It had a nameplate generating capacity of 60 MW but never realized it.

The United Kingdom had a fast reactor program based at Dounreay in the north of Scotland. Its first reactor was a 15-MW experimental fast reactor using sodium-potassium coolant which operated between 1959 and 1977. The reactor initially used uranium metal fuel but later tested uranium oxide fuels. This was followed by a prototype fast reactor with a generating capacity of 250 MW that ran from 1974 until 1994. The latter was a liquid sodium—cooled reactor and was intended to form the basis for a commercial reactor fleet but no commercial versions were built. It used mixed oxide fuel.

France built an experimental fast neutron reactor called Rapsodie in 1967 with no power production capability. The liquid sodium—cooled plant was shut down in 1983. However, soon after the project was started, France also began work on the 250-MW Phénix nuclear fast reactor. The reactor contained just under 1 tonne of plutonium in a fuel that was enriched to 77% plutonium-239. It was connected the grid in 1973 and was finally shut down in 2009. Even as this plant was being built, France, Germany and Italy signed a series of agreements for the construction of two commercial fast neutron reactors, one in Germany and one in France. After a tortuous gestation period work began in 1976 on the first plant, in France and christened Superphénix. The plant finally went critical in 1985 with a rated generating capacity of 1200 MW, the largest such plant ever built. However, the liquid sodium—cooled plant suffered a number of operational problems including a major sodium leak. It was finally shut down in 1998. Superphénix would have formed the basis for a European fast breeder reactor with a generating capacity of 1450 MW but work on that project has virtually been abandoned.

Japan has operated two fast breeder reactors as part of a program to develop fast neutron reactor technology. The first, called Joyo began operating in 1977 and is still functioning as a test bed for fast neutron developments. Alongside this experimental project, Japan also began construction of a prototype fast reactor called Monju. This plant, with a nameplate capacity of 280 MW, achieved criticality in 1994 but experienced a serious sodium leak in 1995 which closed the unit for the next 15 years. It restarted briefly in 2010 but has since been offline. The Japan Atomic Power Company planned a 660-MW demonstration commercial plant, but the project was canceled in the late 1990s following problems with Monju.

The other country with a long history of fast neutron reactors is Russia. Research into the technology has been underway since the 1950s with a small experimental reactor designated BR-5 operating between 1959 and 2004. This was followed by BOR-60, a 12-MW demonstration plant that used uranium oxide fuel enriched to between 45% and 75%. The plant began operating in 1969 and is still active. In 1972, a 350-MW plant, BN-350, was built in Kazakhstan. The liquid-sodium-cooled unit had two sets of heat exchangers outside the reactor vessel. Part of its output was used for desalination. The plant finally closed in 1999.

In 1980, another fast neutron reactor, designated BN-600 with a gross generating capacity of 600 MW, began supplying electricity to the grid. This plant is still operating. The unit is described as a pool-type reactor in which the reactor vessel is immersed in a pool of the liquid sodium coolant as shown in Fig. 6.2. Heat exchangers also filled with liquid sodium carry heat from the pool outside the reactor containment to a heat exchanger/steam generator where it is used to raise steam for steam turbines.

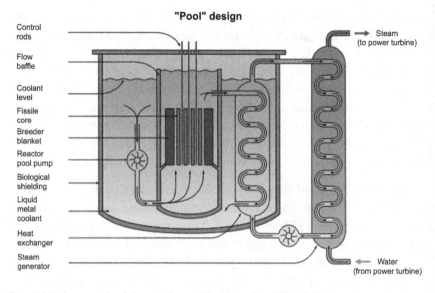

Figure 6.2 Pool-type metal-cooled reactor. Source: Edited image from Wikipedia Commons.

The Russian program has steered away from plutonium-239 and instead uses uranium-235 as the primary fissile material. In the case of the BN-600 the fuel is in the form of uranium oxide which has been enriched to up to 26% with uranium-235. However, this unit is expected to be adapted to burn plutonium from the Russian nuclear weapons stockpile.

The BN-600 has been followed by a larger unit of the same design, the BN800 with a gross generating capacity of 864 MW. It uses oxide fuels with a fissile isotope content of 20–30%, but can also burn uranium and plutonium nitride, or metal fuels too. The plant has primary and secondary cooling loops using sodium to carry heat to the steam generators. A loop-type metal-cooled reactor of this type is shown in Fig. 6.3. Steam is produced at 470°C and the thermal efficiency is claimed to be 39%. Meanwhile the BN-800 is the prototype for a commercial BN-1200 fast neutron reactor with a gross output of 1220 MW.

In addition to its series of liquid-sodium-cooled reactors, Russia has also been exploring reactors that use lead or a lead-bismuth mixture as the coolant. The initial work on these reactors was for submarine propulsion, but more recent designs are intended for civil power generation.

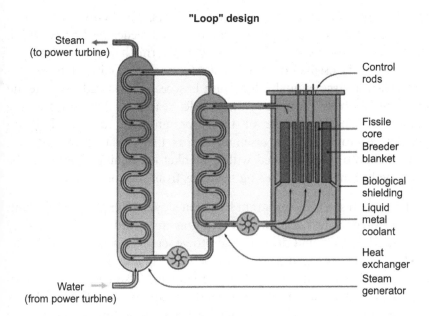

"Loop" design

Steam
(to power turbine)

Control
rods

Fissile
core
Breeder
blanket

Biological
shielding
Liquid
metal
coolant

Heat
exchanger
Steam
generator

Water
(from power turbine)

Figure 6.3 Loop-type metal-cooled reactor. Source: Edited image from Wikipedia Commons.

SLOW NEUTRON BREEDER REACTORS

In addition to plutonium breeder reactors, there is a second type of breeder reactor in which thorium is converted into uranium-233. This isotope of uranium is also fissile, like uranium-235, and so can be used as fuel in a slow neutron reactor. The main interest in this cycle is in India where there are abundant reserves of thorium but much less uranium.

Thorium can be converted into fissile uranium with both slow and fast neutrons. This means that slow neutron reactors such as pressurized water reactors are capable of producing uranium-233 from thorium. In fact pressurized heavy water reactors (PHWRs) are some of the best reactors for this purpose and the Canadian CANDU reactor has been used as a test-bed for thorium conversion into uranium. However the creation of a closed fuel cycle in which more fissile material is generated and then the material reprocessed to make more fuel requires specially designed breeder reactors.

Breeder technology has been under investigation in India since the 1950s and the government has proposed a three-phase strategy leading to a fleet of thorium/uranium breeder reactors. The first stage involves building PHWRs to provide plutonium for a breeder reactor program. In the second stage, this plutonium would form the core of a breeder reactor that is used initially to generate more plutonium but eventually, when a large enough plutonium has been stockpiled, to generate uranium-233 from thorium in the breeder reactor blanket surrounding the core. The third stage of this program then involves building breeder reactors that use uranium-233 as their fissile core materials while the core is surrounded with a blanket containing thorium to be converted to uranium-233 using neutrons from the core.

The progress of this program has been slow. An experimental plant, the fast breeder test reactor (FBTR), was approved in 1971 but only completed in 1985 and did not begin to generate steam until 1993. The FBTR has experienced several accidents including a sodium coolant leak in 2002. Alongside the FBTR, design work also began on a protoype fast breeder reactor with a generating capacity of 500 MW. Work on this began in 2004. It will have a uranium-plutonium oxide core and this will be surrounded by a blanket of thorium, producing fissile uranium-233. Plutonium content in the core will be 21–27%. The unit was supposed to achieve criticality in 2015 but was still not connected to the grid, 1 year later.

CHAPTER 7

Advanced Reactor Design
and Small Modular Reactors

The first commercial nuclear reactors that were built in the 1950s and early 1960s, including early pressurized water reactors (PWRs), boiling water reactors (BWRs), and gas-cooled reactors such as the Magnox reactors are considered today to be first-generation reactors. These were followed by a set of classic, standardized designs that were rolled out commercially in many countries. These are now referred to as second-generation reactors.

These second-generation designs include gas-cooled reactors such as the UK AGR and the Russian RBMK but construction of these gas-cooled lines halted in the early 1980s. By the end of the 1980s, construction of all second-generation reactors had virtually halted as cost spiraled and public opinion turned against nuclear power following the accidents at Three Mile Island in the United States and Chernobyl in Ukraine.

In spite of this virtual moratorium on new nuclear plant construction, all the main nuclear plant manufacturers continued to hope for a nuclear renaissance and a spate of new orders. In order to persuade a skeptical public and utilities that nuclear power remained viable, they all developed new reactor designs that incorporated a host of additional safety features as well as new construction techniques to reduce costs. Many of these designs were originated in the 1980s and offered for construction in the 1990s. All were based on water-cooled reactors but only one of these new designs was built before the end of the 20th century. These reactor designs are now called third-generation reactors.

As well as costs rising dramatically during the 1980s as new safety features were demanded, the approval process for new nuclear plants became extended as a result of the introduction of much stricter environmental and safety requirements. Often each proposed construction project had to go through the same, long procedure even if the new reactor was essentially identical to one already authorized for construction elsewhere. In order to make construction simpler and avoid this massive delay, companies developed standardized designs which they sought to have certified in different regions of the world. Approval by a nuclear certification authority meant that, in principle at least, the planning application for construction of such a plant would be streamlined since the actual reactor design had already been approved. This has now become a key strategy for third-generation reactors.

The certification of third-generation nuclear power plant designs is carried out in the United States by the US Nuclear Regulatory Commission. Since 1997 this organization has certified a number of reactor designs for construction in the United States. In Europe, reactors can be certified for compliance with the European Utility Requirements. A number of third-generation plants have been judged to meet these criteria. In the United Kingdom the Office for Nuclear Regulation undertakes generic design assessment.

Meanwhile nuclear design is progressing further still with a range of fourth-generation nuclear reactor concepts. This effort has been led by the Generation IV International Forum (GIF), a collaborative international effort led by the United States which has identified six reactor concepts for potential future development. Alongside this there are other nuclear technologies being explored including high-temperature gas-cooled reactors and a range of modular small reactors that are intended to be cheap and easy to deploy.

THIRD-GENERATION REACTORS

Third-generation reactors comprise a group of water-cooled reactors that are based on the main branches of second-generation water-cooled reactor design, the PWR, the pressurized heavy water reactor (PHWR), and the BWR. One of the key design improvements in third-generation reactors is the use of passive safety features such that if a reactor goes out of control, naturally driven processes will shut it down. Passive safety means, for example, using natural circulation for the cooling system so that there are no pumps to fail. Other changes are aimed at improving the fuel technology and increasing the overall thermal efficiency of nuclear plants which is low compared to fossil fuel plants. In addition, all third-generation reactors are standardized designs intended to reduce construction costs.

Another feature of third-generation designs, at least for the European market, is the ability to change output rapidly to follow changes in load on the grid. Second-generation reactors were generally designed as base-load power plants intended to operate at close to full output all the time. However the increase in the amount of renewable generation being supplied to grids, especially in Europe, means that many base-load power stations cannot operate at full load all the time because their power is not needed. The European Utility Requirement for new reactors requires that they are able to operate at 25% load and ramp up from 25% to 100% in around 30 minutes. This feature is likely to become important for new reactors on most grids over the coming decade.

Third-generation designs are intended to have an operational life of 60 years, extendable to 120 years with reactor pressure vessel replacement. With improved thermodynamic efficiency, they use their fuel more effectively and the designs aim to burn 17% less uranium for each unit of electricity generated.

The primary third-generation designs were developed during the 1980s but only one was built before the end of the 20th century. That was an advanced boiling water reactor (ABWR) that entered service in Japan in 1996. Since then most companies have refined their designs still further resulting in what are now called Generation III + reactors, denoting that they are third-generation designs but with additional safety and efficiency features.

The evolution of third-generation designs has led to a bewildering array of reactors available from companies across the globe. The main designs are shown in Table 7.1. While the reactors, and their relationships, are complex, most of them can be traced back to one of the main second-generation reactor designs.

One of the most successful new reactors in terms of the number operating is the ABWR, designed by GE and Toshiba and based on the US company's widely used BWR.[1] The design has been certified for use in both the United States and in Europe. As already noted, this was the first third-generation design to be built. Four ABWR reactors are now operating in Japan and two in Taiwan. Others are planned.

[1] GE is now offering the ABWR jointly with Hitachi while Toshiba offers a slightly different version.

Table 7.1 Third-Generation Reactors			
Reactor	Manufacturer	Generating Capacity (MW)	Reactor Type
ABWR	GE-Hitachi, Toshiba	1380	BWR
AP600, AP1000	Westinghouse-Toshiba	1250	PWR
EPR	Areva and EDF	1750	PWR
APR 1400	Korea HNP	1450	PWR
Hualong One	CNNC and CGN	1150	PWR
VVER-1200	Gidropress	1200	PWR
ESBWR	GE-Hitachi	1600	BWR
APWR	Mitsubishi	1530	PWR
Atmea1	Areva and Mistubishi	1150	PWR
EC6	Candu	750	PHWR
ACR1000	Candu	1200	PHWR
VVER-TOI	Gidropress	1300	PWR
Source: *World Nuclear Association, Candu.*			

A Generation III + design that has evolved from the ABWR is the Economic Simplified BWR (ESBWR). The latter has been developed by GE and Hitachi. It further refines the passive elements of the ABWR. The design received certification in the United States in 2014. None has yet been built. There is also a European variant called the EU-ABWR and a UK variant, the UK-ABWR.

While the evolution of the BWR is relatively simple, the heritage of the PWR is complicated by the fact that several different versions appeared around the world. The most direct third-generation descendant of the Westinghouse PWR appears to be the AP600,[2] a 600-MW design that was certified for construction in the United States in 1998. However, no reactors were built to this design. Instead, the company—which is now owned by Japanese firm Toshiba—is promoting a Generation III + evolution of this called the AP1000 with passive features intended to reduce construction costs. The 1250-MW reactor was approved in the United States in 2005. Units based on this design are under construction in China and in the United States. A variant of the AP1000 has also been developed for the Chinese market. Called the CAP1400, it has a generating capacity of 1400 MW.

---

[2] The AP600 also borrows from the System 80 design developed in the 1970s by a company called Combustion Engineering.

China is expected to approve construction of a plant to this design and the Chinese National Nuclear Corp is also hoping to export the technology.

Another branch of the PWR family is the Japanese APWR. Design of this reactor began in the 1980s. It has been developed by five Japanese utilities in collaboration with Mitsubishi and had support from the Japanese government. Its roots can be traced back to Westinghouse, which was involved at an early stage in the design. This reactor is expected to form the basis for the next generation of Japanese PWR reactors though none has yet been constructed. Meanwhile a US variant, the US-APWR has also been developed.

France adopted the PWR design as the basis for its reactor fleet in 1969. The French design is an evolution of the Westinghouse PWR design but is now considered uniquely French. A third-generation design that takes this a stage further is the European Pressurized Reactor (EPR) which has been developed by Areva, Electricité de France, and the German company Siemens. The design builds on both the French and German national PWR evolutions. Four of the reactors are under construction: one in France, one in Finland, and two in China. The two European projects have experienced severe delays and are not expected to enter service before 2018. The Chinese units, where construction started later, may come into service earlier. The original French PWR design is also the starting point for a Chinese reactor called Hualong One.

Yet another reactor with French heritage is Atmea1. This Generation III + design has been developed by Areva and Mitsubishi and incorporates features of both the EPR and the APWR.

A further third-generation lineage can be traced back to a US company called Combustion Engineering (CE). CE built boilers and steam systems for fossil fuel plants and began to supply steam systems for nuclear power plants in the 1960s. Out of this the company evolved its own nuclear reactor design, a PWR to rival the Westinghouse version. This design eventually merged with the Westinghouse design but before that the line led to the development of its own third-generation reactor design called System 80 + . This became the foundation for the Korean APR1400. The APR1400 is based on the Korean Standard Nuclear Power Plant developed by the Korean Electric Power Co. Two APR1400 units are under construction in South Korea and one in the United Arab Emirates. There is a specific European version of this reactor, the APR1400-EUR, which features a double containment and system called a core catcher that will prevent the escape of nuclear material from the core in the event of a melt-down.

The third generation of the CANDU reactor is the ACR series. Two units have been designed, the ACR700 with a generating capacity of 700 MW and the ACR1000 with a capacity of 1200 MW. One of the innovations of the design is to use heavy water in the core but with a light water circuit to extract heat. The reactors would also use uranium enriched to 1.5–2%. The ACR1000 is a Generation III + design. While none has so far been built, some features of the design have been applied to the Enhanced Candu 6 reactor (EC6) (sometimes referred to as the next-generation Candu reactor—see Fig. 7.1), which is an evolution of the Candu 6 reactor design that was built from the 1980s onward in Canada and elsewhere. EC6 is a 700-MW design PHWR that like its predecessors uses natural uranium as its fuel.

The other two third-generation reactors in Table 7.1 are both evolved from the Russian VVER design. The VVER-1200 is the latest evolution of the standard Russian VVER design to meet Generation

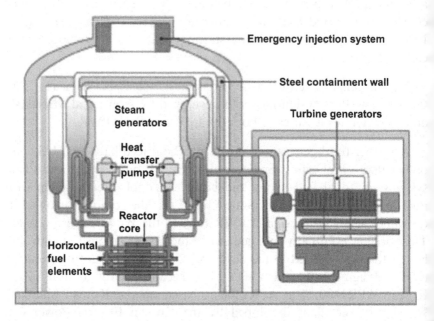

Figure 7.1 Next-generation CANDU reactor. Source: The Institution of Engineering and Technology Nuclear Factsheet.

III + standards. It has a gross generating capacity of 1290 MW. Two units are being built in Russia. A further evolution of this design, called VVER-TOI has also been announced. This has a gross power output of around 1250 MW.

FOURTH-GENERATION REACTOR DESIGNS AND CONCEPTS

While third-generation reactors are beginning to enter service around the world, the industry is already looking beyond these to a new generation of reactors, the fourth generation or Generation IV, which are expected to be ready for commercial rollout from the 2030s onward. One of the key promoters of these new nuclear technologies is the GIF, a collective led by the United States and involving 13 countries including all the world's main nuclear nations. GIF has selected six nuclear technologies that it believes are the best for future development. Three of these six are fast neutron reactors and only two are dedicated slow neutron designs while a third could be either. Most aim to create a closed nuclear cycle where exhausted fuel is reprocessed to provide new fuel and reduce the amount of waste generated by nuclear generation. The different types are considered below.

Gas-cooled fast reactor: One of the six GIF technologies is a gas-cooled reactor concept using helium coolant in a fast neutron reactor design. The nominal operating temperature is 850°C and the reference unit size is 1200 MW. The core would itself provide the breeding element so there would be no surrounding blanket. Fuel is expected to be uranium nitride or carbide. The primary helium cooling cycle would be used to transfer heat to a secondary helium cycle where high-pressure helium would be used to drive a gas turbine. Heat from the gas turbine exhaust would then be captured to raise steam for a steam turbine, a configuration similar to a combined cycle plant.

Sodium-cooled fast reactor: The sodium-cooled fast reactor has been discussed extensively in Chapter 6. The GIF technology would build on existing fast neutron reactor experience and a range of designs are possible including a core and blanket type reactor as well as a reactor in which the core contains the fertile material that breeds new fuel. Both pool-type reactor cores and more conventional cores with cooling circuits passing through them are possible. Like existing fast neutron reactors, the core temperature would be around 550°C and would operate at atmospheric pressure. Unit size could vary from a small 50–150 MW unit to utility-sized plants of up to 1500 MW.

Lead-cooled fast reactor: Lead or lead–bismuth offers an alternative coolant to liquid sodium for fast neutron reactors, as discussed in Chapter 6. Much of the research into the use of the coolant has been carried out in Russia for nuclear submarine propulsion units but there have been recent designs proposed in the United States and Japan too. The envisaged lead-cooled fast neutron reactor would use a uranium metal or nitride fuel as well as being able to burn fissile waste from conventional water-cooled reactors. The initial design is based around a pool reactor in which the core sits in a pool of coolant. The core would operate at 550–800°C. The reactor could be built in a range of sizes from a 1400 MW utility plant down to small "battery" reactors in which the core is provided with enough fuel to operate for 15–20 years without refueling.

Molten salt reactor: The molten salt reactor is primarily a fast neutron reactor concept in which the fissile fuel is dissolved in the molten salt coolant. The salt is sodium fluoride that circulates through a graphite core, the latter providing an element of moderation. The fissile material is continually added to the coolant and waste removed. A range of fissile isotopes can be used as a fuel, including those recycled from conventional reactor waste. The coolant temperature in the reactor is 700–800°C and the core operates at atmospheric pressure. The molten salt forms the primary coolant circuit with the hot salt used to heat water and generate steam for power generation. There is also a variation of the molten salt reactor that uses a uranium/graphite fuel and core with the molten salt acting only as a coolant. This would be a slow neutron reactor, similar to a gas-cooled reactor but with the molten salt instead of gas as the coolant.

Supercritical water-cooled reactor: One of the drawbacks of the conventional water-cooled reactor is that the steam conditions are relatively mild compared to fossil fuel plants and this leads to low thermal efficiency. The supercritical water-cooled reactor is designed to operate at a much higher pressure than a conventional reactor and at higher temperature in order to raise steam above the critical point of water, the point at which the difference between the liquid and gaseous phases disappears. This allows a simpler steam system design and higher efficiency. Typical core temperature would be 500°C. The design is based on the BWR concept in which steam is raised directly within the core and used to drive the steam turbine. In principle a thermal efficiency of 44% would be possible.

Very high-temperature gas-cooled reactor: The very high-temperature gas-cooled reactor is a slow neutron reactor that operates at 900–1000°C. The core is operated under high pressure, it is graphite moderated, and it is cooled using helium. The hot gas exiting the core can be used directly to drive a gas turbine, then heat recovered from the exhaust of the gas turbine to raise steam for a steam turbine. Unit size is generally small, with generating capacities or around 200 MW or less. High-temperature reactors of this type have been explored since the early days of nuclear development but none has yet reached commercial maturity. The structure of the fuel is a key to this type of reactor. The fissile material, which can be uranium dioxide or a uranium/oxygen/carbon compound, is formed into small particles which are coated with graphite and silicon carbide. The particles are then compressed into larger building blocks that are used to create the reactor core. In one design these building blocks are prismatic blocks, in another billiard ball-sized pebbles. Uranium enrichment of up to 20% is envisaged. With the fissile material locked inside a hard shell, the spent fuel is extremely stable and contained. However the future treatment of the spent fuel presents a difficulty that has yet to be overcome.

SMALL MODULAR REACTORS

A number of the designs being developed by GIF, above, include small versions. These form part of a general interest in what have become known as small modular reactors. The small reactor concept has been developed in part to try to meet a perceived need to provide a nuclear option for small grids, particularly in developing countries. The standardized, small size is expected to provide an economical means of providing nuclear power while a modular format means that capacity can be added as demand grows by installing additional modules at the power plant site.

Small modular reactors can be based on any of the nuclear technologies including water-cooled reactors, gas-cooled reactors and some of the novel fourth-generation concepts. Both slow neutron and fast neutron technologies can be adapted to meet the demands of a small size. They are generally defined as being 300 MW or less in generating capacity and may be used for both heat and power production. Designs are expected to be simple with many passive safety features. In addition, most of the components should be capable of being built in a factory and then transported to the site, making the cost of construction much lower and the construction schedule shorter.

Small reactors could offer several unusual features. For example, their small size means that they could be sited underground where they would be shielded from accidents caused by external impacts such as an aircraft crash and isolated in case of an internal accident, so that radiation would not be released into the environment. Another proposal is to build "battery" type reactors that contain sufficient nuclear fuel to operate for 10−20 years without refueling.

Interest in these low-capacity reactors can be traced back to the early days of nuclear power development but few have been built except as experimental units. There are a small number operating today. One of the oldest is the Russian EGP-6, a heat and power reactor based on a graphite moderated, water-cooled design that has a rating of 62 MW_{th} and an electrical output of 11 MW of power. Four of these units, which are essentially scaled-down version of the RBMK reactor, have operated since 1976 at Bilibino in Siberia.

Two examples of a small Chinese PWR reactor called the CNP-300 are currently in service, one in Pakistan and a second in China. The generating capacity is 320 MW. In India, meanwhile, a number of small versions of the Canadian Candu PHWR are operating. The earliest of these entered service in 1984 with a generating capacity of 170 MW. More recent versions have an output of 220 MW.

The most important new small modular reactor development is the construction of two 105 MW high-temperature gas-cooled reactors in China. These are based on a pebble bed reactor design which uses fissile fuel embedded into graphite spheres, as shown in Fig. 7.2. The fuel is enriched to 8.5%. The core is cooled using helium which exits the core at 750°C and is used to raise steam at 566°C.

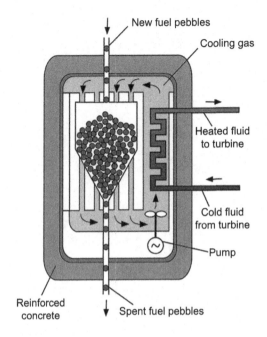

New fuel pebbles

Cooling gas

Heated fluid
to turbine

Cold fluid
from turbine

Pump

Reinforced
concrete

Spent fuel pebbles

Figure 7.2 Pebble bed reactor. Source: Wikipedia commons.

Designs for another 10 or more small modular reactors have reached an advanced stage of development around the world. Many of these are small PWR reactors but there are also small versions of fast neutron reactors too. Most of the research is taking place in the United States, Russia and China, with one project in South Korea.

CHAPTER 8

Nuclear Fission

Nuclear fusion, the reaction that fuels the Sun and the stars, has excited scientists and technologists ever since the process was identified during the 1930s. Similar to fission, the key to unraveling the mechanism of fusion depends on recognizing that some elements in the periodic table are more stable than others. In particular, elements toward the middle of the periodic table become relatively more stable than those at the extremes. Thus, helium with two protons and two neutrons is more stable than hydrogen with one proton. Identifying this during the early part of the last century culminated in the realization by a British astrophysicist, Sir Arthur Eddington, that fusion of hydrogen atoms to produce helium would release around 0.7% of their mass as energy. This led to the first theory of fusion in the stars, elaborated by German physicist Hans Bethe in 1939.

Unsuccessful attempts at fusion took place during the 1930s at the Cavendish Laboratory in Cambridge, UK, but work on fusion halted during World War II. Experimental work restarted during the late 1940s, again in the United Kingdom, at Harwell and it was here that the Zero Energy Toroidal Assembly or ZETA operated between 1954 and 1958 and provided a body of experimental observation that proved vital for later fusion research. Since then a series of fusion reactors have been built around the world. Around 20 are in operation today.

In 1958, at an *Atoms for Peace conference* in Geneva, fusion research was established as an international collaborative venture and at least one strand of fusion development, that based on magnetic confinement, has remained international in flavor ever since. The necessity for this was reinforced during the 1970s when it became clear that the cost of developing fusion was likely to be beyond the resources of any one nation. Before that a key magnetic confinement discovery was made by Russian scientists Igor Tamm and Addrei Sakharov who, in 1968, unveiled a magnetic confinement device called a tokamak that was capable of sustaining much higher temperatures than any previous device. Although other systems are still being explored, this is now the dominant method of using a magnetic field to contain a fusion plasma.

The latest large reactor exploiting this technology is the International Thermonuclear Experimental Reactor (ITER) under construction in the south of France. If successful when it reaches full power, by 2030 or earlier, it will be the first fusion reactor to produce more energy that is needed to start the reaction in the first place. However a commercial fusion reactor is not likely before the middle of the century, at the earliest.

While large fusion reactors based on magnetic confinement were being built in the United States, Europe, and Japan, other fusion developments remained hidden behind the security of nuclear armaments research. Fusion is the basis for the hydrogen bomb and so much of the research into its development and control remained secret until very recently. It is this research that has led to as second approach to generating electrical power from fusion, the idea of inertial confinement. During the last 5 years, the veil of secrecy has at least partly dropped from this research and a major program in the United States aims to develop a demonstration power plant based on these fusion techniques during the 2020s. If successful, it may reach fruition at around the same time as ITER.

FUSION BASICS

Nuclear fusion, like nuclear fission, can provide energy from mass. In the case of fusion, this energy is released when very light atoms are turned into slightly heavier, but more stable atoms. Since the primary precursor is hydrogen, one of the most abundant elements on the earth, fusion—in principle at least—offers a limitless supply of energy. Moreover the fusion reaction, while producing a large amount of energy, generates less toxic waste than nuclear fission. The reaction is not entirely clean because very high-energy particles are generated and these will cause nuclear reactions in plant components, leaving some radioactive remnants. One of the key hydrogen isotopes involved in fusion is radioactive too. Nevertheless, it is generally judged much more benign, environmentally than fusion.

The fusion reaction that powers the Sun and stars is a reaction in which hydrogen atoms combine to produce deuterium and then deuterium and hydrogen atoms fuse to make helium with the release of energy. This reaction takes place in the center of the Sun at a temperature of 10 million to 15 million degrees celsius and under extreme pressure. Under these conditions, the hydrogen atoms disintegrate to form a sea of electrons and nuclei, which are held close together by the massive gravitational force within the Sun (gravitational confinement). The conditions required to allow this reaction to take place are considered almost impossible to recreate on the necessary scale on Earth. However, there is another fusion reaction, between deuterium and a third isotope of hydrogen called tritium, that requires less extreme conditions and these can be recreated, albeit with extreme difficulty, on our planet. It is this reaction that forms the basis for fusion research.

As with fission, this fusion reaction releases its energy primarily as kinetic energy that is carried away by a neutron that is generated during the reaction. In a fusion reactor this energy must be captured and used to generate steam for power production. The amount of energy available is enormous. In theory 1 tonne of deuterium could provide the equivalent of 3×10^{10} tonnes of coal.

There is one problem with the deuterium—tritium reaction; tritium does not occur naturally and must itself be made during a nuclear reaction. It can be produced from lithium using the high-energy neutrons in the fusion reactor so like a breeder reactor, a fusion reactor will have to be able to produce its own fuel as well as energy. This significantly complicates the design of such reactors.

MAGNETIC CONFINEMENT

The fusion reaction between deuterium and tritium (the DT reaction) discussed above is the easiest to achieve in a reactor but even so it requires extremes of both temperature and pressure. The temperature required to achieve fusion with DT is over 40 million degrees celsius. This temperature is necessary to provide the nuclei with enough thermal energy to overcome the electrostatic repulsion between the two positively charged particles and allow them to approach close enough to one another to react. However at temperatures anywhere near this, atoms disintegrate to create a sea of electrons and nuclei, a fourth state of matter called a plasma.

If enough energy can be provided to allow the DT to enter the plasma state and for fusion to begin, then in principle the reaction itself will be self-sustaining because the enormous amount of energy released by each reaction will maintain the high temperature. However the problem of achieving this is complicated by the fact that the plasma must be maintained for a sufficiently long time for the reaction to take place, and the density of the plasma[1] must be high enough so that it can produce more energy than it absorbs.

There are no materials in existence that can be used to build a container to hold a plasma so an alternative way has to be found to contain and control the hot sea of particles. The most promising solution is by means of a magnetic field. If a strong magnetic field is applied to the plasma, then the motion of the charged particles in the plasma will be affected by this field. By judicious application of magnetic fields, the charged particles can be made to orbit inside the reactor without touching its walls. This is the basis of magnetic confinement.

Magnetic confinement was recognized early in fusion research as the only way to maintain a fusion plasma but it was not until the 1950s that the best form of magnetic field, the toroidal field, was identified by scientists in Russia. It was there that a device called a tokamak was developed and in the early 1960s experimental results showed that the high temperatures required for fusion could be achieved with this device.

Since then, a series of ever larger tokamak reactors have been constructed. The most important of these were the Joint European Torus (JET) at Culham in the United Kingdom, the Tokomak Fusion Test Reactor (TFTR) at Princeton in the United States, JT-60U in Naka, Japan, and T-15 in Moscow, Russia. Both TFTR and JET experimented with DT fuel from the beginning of the 1990s, and in 1997 JET established the record for the greatest amount of energy generated by a fusion reactor, 16 MW. A schematic of a magnetic confinement fusion plant is shown in Fig. 8.1.

Even with this output, the reactor consumed more energy than it generated. JET achieved a power-in-to-power-out ratio (the gain of the reactor) around 0.7. A gain of 1 represents the break-even point. Even more crucially, JET could only maintain the plasma burst for 5 seconds. If it continued for longer, its systems would begin to overheat.

---

[1]In fact it is the ion density that is important.

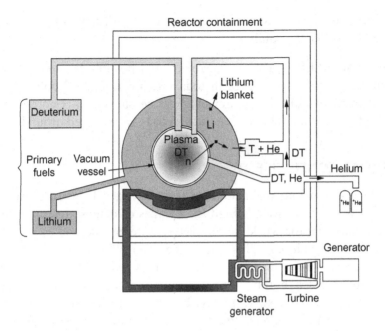

Figure 8.1 Magnetic confinement fusion power plant schematic.

During the experiment, the temperature at the center of the JET plasma reached 170 million degrees celsius. At this temperature the plasma behaves like a boiling liquid, creating eddy currents that make is unstable. Controlling turbulence within the plasma is one of the keys to building an efficient fusion reactor that can operate with a gain of greater than one.

The next stage in magnetic confinement fusion development is the International Thermonuclear Experimental Reactor (ITER – the word is also Latin for "the way"). This project was first conceived in 1988 and an agreement to build the plant was finally signed in 2007 by the EU, Russia, Japan, the United States, China, India, and South Korea.

ITER will have a plasma volume of 800 m^3 and a power output of 500 MW$_{th}$, 30 × that of JET. Fusion energy generation is a matter of size and at this size it is hoped that ITER will have a gain of 10, producing 500 MW$_{th}$ from an input of 50 MW$_{th}$. This will be sufficient to prove fusion as a net source of energy but ITER has not been designed to generate power so it will not have all the features needed for a demonstration plant. Its main goal is to generate 500 MW of thermal fusion power for 400 seconds to prove that commercial fusion is possible. The first demonstration power plant will have to wait until a successor to ITER is built.

Even so, ITER is a massive project, and possibly the most difficult engineering project on the earth today. It involves the 7 collaborating nation members and the components will be built in 34 countries. If all proceeds according to plan, then the first tests are expected around 2020.

INERTIAL CONFINEMENT

As the research into magnetic confinement edged forward during the past 60 years, there was, behind the facades of defense research establishments across the world, a completely different method of achieving fusion being explored. This had more interest in how to create an instantaneous fusion reaction to release an explosive amount of energy, in other worlds a bomb. Secrecy meant that this research remained hidden until very recently. Today, however, the commercial potential of this research is being explored and much more knowledge about the technique called inertial confinement is available publicly.

While magnetic confinement seeks to create a stable continuous plasma in which fusion can take place the alternative, inertial confinement, seeks instead to generate energy from a series of discrete fusion reactions, each producing a burst of energy. In an inertial confinement reactor, small capsules containing around 150 mg of a mixture of deuterium and hydrogen are exposed to a massive pulse of energy from multiple lasers. When the laser beams strike the capsule, they create an explosion of X-rays from its surface and these in turn (by the mechanical principle of action and reaction) create a pressure pulse—a shock wave—that heats and compresses the DT mixture within the capsule with such vigor that the conditions for fusion are generated at its core. One fusion starts, the reaction radiates outward through the DT mixture faster than the actual molecules can expand and escape (they are "confined" by their inertia) and so the whole charge undergoes fusion and releases a pulse of energy. The inertial confinement fusion process is shown schematically in Fig. 8.2.

To make this into a means of generating power, these small exploding Suns must be created at a relatively rapid rate of perhaps 15 each second. This sounds both exacting and ambitious but it is exactly what a US program proposes. To achieve it, the US government has built

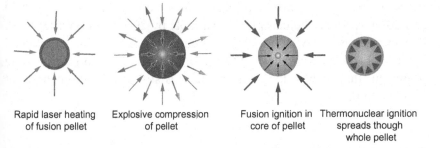

| Rapid laser heating of fusion pellet | Explosive compression of pellet | Fusion ignition in core of pellet | Thermonuclear ignition spreads though whole pellet |

Figure 8.2 The principle of inertial confinement.

the National Ignition Facility (NIF), a $5 billion project that is intended to serve both military and civilian research.[2]

NIF is provided with 192 lasers capable of providing up to 5 MJ of energy in a 20-nanosecond pulse and it has so far produced 1.8 GJ in a single short pulse, equivalent to a delivery rate of 500 TW of energy. The lasers produce infrared light which is converted first into visible light and then into ultraviolet before striking its target. All the energy contained in the laser beams is focused onto 150 mg of DT inside a tiny capsule, about 2 mm in diameter, called a hohlraum.

NIF is only capable of single shot experiments rather than the continuous operation required for a power plant. However the amount of laser power it has at its disposal in similar to that which would be required for a 1000 MW power station. Since it started in 2009, it has carried out a series of experiments trying to achieve ignition, the point at which the DT produces more fusion energy than the lasers pump into it. By 2013 it was a factor or two or three short of ignition.

Running alongside NIF is another project, the Laser Inertial Fusion Energy (LIFE) project. LIFE is a collaboration of scientists, technologists, utilities, and regulators that are seeking to design a power plant capable of exploiting inertial confinement. The target for LIFE is to be in a position to build a demonstration power plant within 10 years of ignition being achieved at NIF. This demonstration plant would have an initial generating capacity of 400 MW but would be designed to be capable of scaling up to 1000 MW. Current plans

The National Ignition Facility was built in response to the Comprehensive Test Ban Treaty that outlaws nuclear weapons testing. It will be used to support the treaty as well as providing research into inertial fusion for power generation.

envisage this demonstration project will be constructed between 2020 and 2030 with commercial plants available by 2030 or soon afterward. However, it is a saying in fusion research circles that a commercial plant is always 30 years away. While the future for fusion looks more promising than it has done before, there is still an enormous amount of work to be done.

TRITIUM PRODUCTION

For either magnetic confinement or inertial confinement to become a successful source of electrical power, each needs a source of tritium. Tritium is radioactive with a half-life of 10 years so any naturally formed tritium quickly decays. However a supply can be generated in a nuclear reaction from lithium. When lithium is exposed to neutrons it splits for form an atom of tritium and an atom of helium. One isotope of lithium, lithium-6 will react with slow neutrons while another isotope, lithium-7 reacts with fast neutrons.

In order to provide self-sustaining power plants, both a magnetic confinement and an inertial confinement plant would have to produce, or breed, its own tritium. Under current designs, it is conceived that this will be manufactured in a "blanket" that surrounds the core of the fusion reactor, or be placed around the ignition chamber in an inertial confinement plant. The design of the blanket for either type of reactor is still a matter of speculation. However, it will have to serve two functions. The first is to absorb the high-energy neutrons that are generated during the fusion reaction, capturing their heat in the process. The second is to generate tritium for the reactor.

How lithium will be utilized within a plant is uncertain although several options exist. One is to use lithium as the plant coolant as well as the tritium breeder. If used in this way, liquid lithium would circulate in primary coolant circuits, capturing heat and carrying it to a heat exchanger and steam generator where steam is produced to drive a steam turbine generator. Tritium that is generated in the blanket will dissolve in the hot lithium and would then have to be harvested. Various schemes for this have been proposed but so far most are only experimental. An alternative approach is to have the lithium present in ceramic blocks through which a coolant such as helium passes. This would separate the cooling system from the tritium breeding system.

but would require the ceramic blocks to be removed to harvest the tritium from them.

Safety and waste disposal are also issues that have to be addressed in both types of fusion plant. As already noted, the fusion reaction produces high-energy neutrons and these will cause nuclear reactions within the components of the power plant, creating a radioactive waste problem. The main source of this waste is likely to be the steel that is used to construct the reactor, of either type. There are low activation steels available that will produce short-lived radioactive isotopes when exposed to neutrons. These then decay quickly so that the level of radioactivity falls rapidly once the reactor is decommissioned and the steel can be consigned to low-level radioactive waste repositories.

Another consideration is the tritium. This hydrogen isotope that plays a key role in the fusion reaction is radioactive, though again with a short half-life so that it will decay rapidly. A fusion reactor is expected to contain less than 1 kg of tritium while it is operating. Further, a plant failure of any type is likely to lead to the fusion reaction shutting down since the reaction is not self-sustaining, reducing the overall risk level. Even so it presents an issue for regulators if, and when fusion plants become commercial.

CHAPTER 9

The Environment Effects of Nuclear Power

The use of nuclear power raises important environmental questions and as with most environmental issues it is a matter of weighing advantages and disadvantages. The positive aspects of nuclear power include the fact that generating electricity from a nuclear station does not involve the release of any carbon dioxide into the atmosphere and so a nuclear plant can provide part of the solution to reducing global warming. Nuclear power can also provide energy security in countries that have limited natural energy resources. The negative aspects of nuclear power are all linked to the fact that nuclear generation is based on exploitation of nuclear reactions. These reactions produce a range of potentially hazardous waste products. In addition, the effects of a major accident at a nuclear power plant can be far-reaching. When there have been accidents these have had a massive effect on the popular perception of nuclear power.

Added to these considerations is another that the nuclear industry has to address, the perceived link between nuclear and nuclear weapons. While the nuclear industry would claims that the civilian use of nuclear power is a separate issue to that of atomic weapons, the situation is not that clear cut. Nuclear reactors are the source of plutonium which can be used to make a nuclear weapon. Plutonium creation depends on the reactor design and it is possible to build nuclear reactors that produce very little or no nuclear isotopes that are useful for weapons production. However the reactors that are in use today virtually all produce material that can be used for weapons. In addition, most nuclear power plants require enriched uranium and therefore rely on uranium enrichment plants. Highly enriched uranium is another material capable of being fashioned into a bomb. Both are therefore areas of international concern.

The danger is widely recognized. Part of the role of the International Atomic Energy Agency is to monitor nuclear reactors and track their inventories of nuclear material to ensure than none is being sidetracked into nuclear weapons construction. Unfortunately, this system can never be foolproof. It seems that only if all nations can be persuaded to abandon nuclear weapons can this danger be removed. Such an agreement looks highly improbable.

The effects of the detonation of a nuclear weapon are devastating, as history has clearly demonstrated. Of course a nuclear power plant is not a nuclear bomb. Unfortunately for the nuclear power industry, some of the after-effects of the detonation of a nuclear device can also be produced by a major civilian nuclear accident. The contents of a nuclear reactor core includes significant quantities of extremely radioactive nuclei. If these were released during a nuclear accident they would almost inevitably find their way into humans and animals via the atmosphere or through the food chain.

Large doses of radioactivity or exposure to large quantities of radioactive material kills relatively swiftly. Smaller quantities of radioactive material are lethal too, but over longer time scales. The most insidious effect is the genesis of a wide variety of cancers, many of which may not become apparent for 20 years or more. Other effects include genetic mutation which can lead to birth defects.

The prospect of an accident leading to a major release of radioactive nuclides has created a great deal of popular apprehension about nuclear power. The industry has gone to extreme lengths to tackle this apprehension by building ever more sophisticated safety features into their power plants. Unfortunately the accidents at Three Mile Island in the United States, Chernobyl in Ukraine, and Fukushima Daiichi in Japan suggest that it may be impossible to build a nuclear power plant that is entirely safe. For modern plants the risk of an accident may be extremely low. The difficulty is in persuading the public that any risk is acceptable when the stakes are so high.

Unfortunately the fear associated with nuclear accidents has recently been magnified by a rise in international terrorism. The threat now exists that a terrorist organization might seek to cause a nuclear power plant accident or, by exploiting contraband radioactive waste or fissile material, cause widespread nuclear contamination.

So far a peacetime nuclear incident of catastrophic proportions has been avoided, though both Chernobyl and Fukushima caused extensive disruption and in the case of the former a disputed number of deaths as a result of radioactive exposure. Smaller incidents have been more common and low-level releases of radioactive material have taken place. While these are rarely serious, they raise other issues.

One of these issues is the level of the danger from exposure to low radiation levels. The effects of low levels of radioactivity have proved difficult to quantify. Safe exposure levels are used by industry and regulators but these have been widely disputed. On one hand, some would claim that there is no safe level of exposure. On the other hand there are natural sources of radiation to which everybody on the planet is exposed, so a level of exposure that is lower than that experienced naturally might be considered insignificant. Again, it is a matter of trying to establish risk levels, and then to determine what level of risk is acceptable.

NUCLEAR POWER AND GLOBAL WARMING

One of the main advantages of nuclear power promoted by the nuclear industry today relates to its ability to provide low carbon electricity generation. According to the International Energy Agency (IEA), nuclear power was the largest source of low carbon electricity among the countries of the Organization for Economic Cooperation and Development (OECD) countries in 2013 with 18% of total electricity production. Across the globe as a whole its share of production was 11%, making it the second largest contributor after hydropower.[1]

Based on an IEA scenario for future power generation under which the rise in the global temperature is restricted to 2°C, the organization suggests that nuclear generation would need to more than double from its present level, reaching 930 GW of installed capacity by 2050, when it would provide roughly 17% of global electricity production. This would represent an ambitious program of nuclear construction. However it faces a number of hurdles.

Much of the existing nuclear capacity is in OECD countries. The members of the OECD are mostly rich, developed nations, and many of these invested in nuclear power in the early days of nuclear evolution. As a consequence, many of these countries have benefited

[1] Technology Roadmap: Nuclear Energy, 2015 Edition, IEA and NEA, 2015.

from fleets of nuclear plants providing cheap base-load power. The cost of new nuclear power plants makes the technology inaccessible to all but the richest nations today, yet growth in generating capacity is more urgent elsewhere. Small modular nuclear power plants might make the technology more accessible but such plants are not currently available commercially.

On top of that, the recent nuclear accident in Japan has curtailed much global nuclear activity while the construction of renewable generating capacity from wind and solar power continues to grow. There is already evidence that the low cost of these technologies is beginning to undercut others, particularly nuclear power, and this trend will continue. Meanwhile the challenging financial situation across the globe in the second decade of the 21st century makes it extremely difficult to find funding for capital intensive projects like nuclear power stations.

So, while nuclear power can contribute to reducing global warming, it will not be the first choice for new generating capacity in many, if not most parts of the world. Perceptions may change and if the cost of nuclear construction can be reduced by the availability of small, standardized nuclear power units then the appetite for nuclear power may improve. The danger is that nuclear technology will be overtaken by renewable developments elsewhere and that when new nuclear technologies are available, nuclear growth will be difficult to justify economically.

RADIOACTIVE WASTE

As the uranium fuel within a nuclear reactor undergoes fission, it generates a cocktail of radioactive atoms within the fuel pellets. Eventually the fissile uranium becomes of too low a concentration to sustain a nuclear reaction. At this point the fuel rod will be removed from the reactor. It must now be disposed of in a safe manner. Yet after more than 60 years of nuclear fission, no safe method of disposal is widely available.

Several options are considered viable. The best large-scale method would appear to be disposal of waste in underground bunkers built in stable rock structures. However while the principle has been agreed finding a site where construction can take place has proved extremely difficult. Reprocessing the waste fuel to remove and reuse the uranium and plutonium fissile material it contains another option. This would reduce the volume of the residual waste, though the residue of high-level waste still requires secure disposal. Spent fuel reprocessing has been carried out in one or two countries but in most there is no agreed solution. As a result, most spent nuclear fuel has been stored in ponds at the nuclear power plants where it was produced. This is now causing its own problems as storage ponds designed to store a few years' waste become filled, or overflowing.

Meanwhile, radioactive waste disposal has become one of the key environmental battlegrounds over which the future of nuclear power has been fought. Environmentalists argue that no system of waste disposal can be absolutely safe, either now or in the future. And since some radioactive nuclides will remain a danger for thousands of years, the future is an important consideration.

The quest to solve the problem continues. For many years underground burial has been the preferred option for the nuclear industry. This requires both a means to encapsulate the waste and a place to store the waste once encapsulated. Encapsulation techniques have already been developed. These include sealing the waste in a glass-like matrix that is then stored in heavy steel containers. The waste still generates heat, even in this form, and so must be cooled once it has been encapsulated. However the encapsulation should make it impossible for the waste to escape into the environment.

While encapsulation has been demonstrated, construction of an underground store has yet to be realized. An underground site must be in stable rock formation in a region not subject to seismic disturbance. Such locations have been identified and sites in the United States and Europe have been studied for many years but none has yet been built. The most advanced project of this type is the Onkalo repository in Finland where excavation of the underground cavern has begun but the project awaits a construction license from the government. If this is granted, the first waste fuel is expected to be stored around 2020. However this site is designed for waste from Finnish nuclear plants. Other countries still have to find their own solutions.

Fuel reprocessing may offer another, partial solution to the problem of nuclear waste. The reprocessing of spent fuel is part of the nuclear fuel cycle, enabling fissile uranium and plutonium to be recovered from the fuel waste and reused in nuclear fuel. However once reprocessing is complete there is still a significant residue that contains a variety of radioactive isotopes that are of no use in reactors. So while reprocessing reduces the volume of waste, it does not entirely solve the problem.

Other schemes have been proposed for nuclear waste disposal. It is possible to return the high-level waste containing radioactive isotopes to a reactor where they are bombarded with neutrons and where they eventually react to produce less harmful isotopes. This appears to be costly. Another involves loading the fuel into a rocket and shooting it into the sun. Yet another proposes utilizing particle accelerators to destroy the radioactive material generated during fission.

Unfortunately, while there are many proposals for the disposals of radioactive waste, there are limited practical solutions available. In the meantime the volume of radioactive waste continues to increase and so does the environmental problem it represents.

WASTE CATEGORIES

Spent nuclear fuel and the waste from reprocessing plants represent the most dangerous of radioactive wastes but there are other types too. These come from a variety of sources. Anything within a nuclear power plant that has even the smallest expose to any radioactive material must be considered contaminated. One of the greatest sources of such waste is the fabric of a nuclear power plant itself This creates a large volume of waste when a nuclear power plant is decommissioned.

High-level wastes are expected to remain radioactive for thousands of years. It is these wastes which cause the greatest concern and for which some storage or disposal solution is most urgently required. But these wastes form a very small part of the nuclear waste generated by the industry. According to the World Nuclear Association the high-level waste only makes up around 3% of the total by volume Most is low-level waste. Even so it too must be disposed of safely. In order to deal with these different wastes, regulatory authorities have developed nuclear waste categories.

In the US spend fuel and the residual waste from reprocessing plants is categorized as high-level waste[2] while reminder of the waste from nuclear power plant operations is classified as low-level waste. There is also a category called transuranic waste which is waste containing traces of elements with atomic numbers greater than that of uranium (92). All elements with a higher atomic number than uranium are naturally radioactive. Low-level wastes are further subdivided into classes depending on the amount of radioactivity per unit volume they contain.

In the United Kingdom there are three categories of waste, high level, intermediate level, and low level. High level includes spent fuel and reprocessing plant waste, intermediate level is mainly the metal cases from fuel rods and low-level waste constitutes the remainder. Normally both high and intermediate level waste require some form of screening to protect workers while low-level waste can be handled without a protective radioactive screen.

Low-level waste will often disposed of by shallow burial, often after compacting, and in some cases it may be incinerated is a special waste combustion plant to reduce the volume before burial. Intermediate level waste needs has to be shielded since it contains higher levels of radioactivity. It may be sealed in concrete of bitumen before burial. However, unlike high-level waste, this type of waste does not need cooling when it is stored.

DECOMMISSIONING

A nuclear power plant will eventually reach the end of its life and when it does it must be decommissioned. At this stage the final, and perhaps largest nuclear waste problem arises. After 30 or more years[3] of generating power from nuclear fission, most of the components of the plant have become contaminated and must be treated as radioactive waste. This presents a problem that is enormous in scale and costly in both manpower and financial terms.

[2] The US Department of Energy does not classify spent fuel as waste but the Nuclear Regulatory Commission does.
[3] Nuclear plants in many parts of the world are now seeking operating license extensions to allow them to continue operations for up to 60 years.

The cleanest solution is to completely dismantle the plant and dispose of the radioactive debris safely. This is also the most expensive option. A half-way solution is to remove the most radioactive components and then seal up the plant from 20 to 50 years, allowing the low-level waste to decay, before tackling the rest. Two Magnox reactor buildings in the United Kingdom were sealed in this way in 2011 and are expected to remain in that state for 65 years. A third solution is to seal the plant up with everything inside and leave it, entombed, for hundreds of years. This has been the fate of the Chernobyl plant.

Decommissioning is a costly process. Regulations in many countries now require that a nuclear generating company put by sufficient funds to pay for decommissioning of its plants. The US Nuclear Energy Institute suggests that the cost of decommissioning a US power plant is between $450 million and $1.3 billion, based on figures from the US Nuclear Regulatory Commission from 2013. The US utility Southern California Edison has put aside $2.7 billion to decommission its San Onofre power plant, expecting this to cover around 90% of the total expenditure. Meanwhile in 2011 the UK government estimated nuclear decommissioning costs for its existing power plants to be £54 billion. When building a new nuclear plant, the cost of decommissioning must therefore be taken into account.

NORMAL POWER PLANT ENVIRONMENTAL EFFECTS

Aside from the nuclear aspect of a nuclear power plant there are other environmental effects which a nuclear station will share with other types of power station. There will be some environmental disruption while the plant is being constructed and significant additional vehicle movements; for a nuclear power plant this disruption will last for several years. There will be some habitat destruction, some air pollution and noise which will affect any communities or habitations in the vicinity. There will also be disruption associated with the connection of the power plant to the grid which may require a lengthy right-of-way and transmission line construction.

Once the plant starts operating there will be less vehicle movement and operation of the plant itself will be relatively quiet. However most nuclear power plants need water for cooling the steam turbine condenser and this will require the pumping of water from a local source and then returning it, but at a higher temperature. This is likely to cause changes to the aquatic or marine environment depending on whether fresh or salt water is being used for cooling.

A nuclear power plant is likely to have a long service life but when it does reach the end of its life, the process of decommissioning the plant will be long and complex and this will again involve high levels of activity and vehicle movement. Some of the activities associated with construction, operation, and decommissioning of the station will be disruptive but these are activities that would be associated with virtually any power plant construction.

CHAPTER 10

The Cost of Electricity From Nuclear Power Stations

The cost of electricity from a power plant of any type depends on a range of factors. First there is the cost of building the power station and buying all the components needed for its construction. In addition, most large power projects today are financed using loans so there will also be a cost associated with paying back the loan, with interest. Then there is the cost of operating and maintaining the plant over its lifetime, including fuel costs. Finally the overall cost equation should include the cost of decommissioning the power station once it is removed from service.

It would be possible to add up all these cost elements to provide a total cost of building and running the power station over its lifetime, including the cost of decommissioning, and then dividing this total by the total number of units of electricity that the power station produced over its lifetime. The result would be the real lifetime cost of electricity from the plant. Unfortunately such a calculation could only be completed once the power station was no longer in service. From a practical point of view, this would not be of much use. The point in time at which the cost-of-electricity calculation of this type is most needed is before the power station is built. This is when a decision is made to build a particular type of power plant, based normally on the technology that will offer the least cost electricity over its lifetime.

LEVELIZED COST OF ENERGY MODEL

In order to get around this problem, economists have devised a model that provides an estimate of the lifetime cost of electricity before the station is built. Of course, since the plant does not yet exist, the model requires a large number of assumptions to be made. In order to make this model as useful as possible, all future costs are also converted to the equivalent cost today by using a parameter known as the discount rate. The discount rate is almost the same as the interest rate and relates to the way in which the value of one unit of currency falls (most usually, but it could rise) in the future. This allows, for example, the maintenance cost of a nuclear steam turbine 20 years into the future to be converted into an equivalent cost today. The discount rate can also be applied the cost of electricity from the nuclear plant in 20 years time.

The economic model is called the levelized cost of electricity (LCOE) model. It contains a lot of assumptions and flaws, but it is the most commonly used method available for estimating the cost of electricity from a new power plant.

When considering the economics of new power plants the levelized cost is one factor to consider. Another is the overall capital cost of building the generating facility. This has a significant effect on the cost of electricity, but it is also important because it shows the financial investment that will have to be made before the power plant generates any electricity. The comparative size of the investment needed to build different types of power stations may determine the actual type of plant built, even before the cost of electricity is taken into account. This is of importance with nuclear power plants as their capital cost is among the highest of all generating technologies. Capital cost is usually expressed in terms of the cost per kilowatt of generating capacity to allow comparisons between technologies to me made.

When comparing different types of power station, there are other factors that need to be considered too. The type of fuel, if any, that is used is one. A coal-fired power station costs much more to build than a gas-fired power station, but the fuel it burns is relatively cheap. Natural gas is more expensive than coal and it has historically shown much greater price volatility than coal. This means that while the gas-fired station may require lower initial investment, it might prove more expensive to operate in the future if gas prices rise dramatically.

Renewable power plants can also be relatively expensive to build. However, they normally have no fuel costs because the energy they exploit is from a river, from the wind, or from the Sun and there is no economic cost for taking that energy. That means that once the renewable power plant has been paid for, the electricity it produces will have a very low cost. All these factors may need to be balanced when making a decision to build a new power station.

For a nuclear station, the cost of fuel is normally considered to be low compared to fossil fuel plants. Estimates vary, but according to the US Nuclear Energy Institute the cost of fuel accounts for around 28% of the cost of electricity from a nuclear power plant compared to 78% for a coal-fired power plant and 89% for a gas-fired power plant. This means that any volatility in nuclear fuel prices will have a smaller effect on the cost of electricity than it would in a gas-fired power plant.

CAPITAL COST

The capital cost of a nuclear power plant makes one of the greatest contributions to the overall cost of electricity from nuclear power. Nuclear power plants are large and complex projects that will usually take a long time to build. They use a range of high-technology materials including steel for the reactor vessel and more exotic materials in nuclear-specific parts of the plant such as control rods or fuel casings. There will also be a large amount of concrete. This means that not only is a project expensive because of its scale, but also that it may be vulnerable to changes in commodity prices that could affect overall cost during construction.

For the latest third-generation projects, there is also the danger of cost overruns resulting from the fact that early versions of these plants are essentially prototypes based on designs that have never been built before. Two European pressurized reactors under construction, one in France and other in Finland, have experienced lengthy time and cost overruns as a consequence of problems during construction. In 2015, the cost of the Finnish plant was estimated to have risen from an initial €3.2 billion to €8.5 billion with similar cost escalation for the French plant.

Table 10.1 contains figures for the overnight cost[1] of nuclear power plants in the United States based on figures from the US Energy Information Administration (EIA). The figures are from the EIA's Annual Energy Outlook series and refer to the cost of a plant in the year previous to the date of the report.

The overnight cost is the basic cost of building the power plant without any financing costs being taken into consideration.

Table 10.1 Overnight Capital Cost of Nuclear Power Plants	
Year	Overnight Capital Cost ($ per kW)
2001	1729
2003	1750
2005	1694
2007	1802
2009	2874
2011	4567
2013	4700
2015	4646
Source: *US Energy Information Administration.*[2]	

The 2001 report estimated the overnight cost of a nuclear power station in 2000 to be $1729 per kW. Costs remained relatively static for the following 6 years, so that the 2007 report estimated the cost in 2006 to be $1802 per kW. By the time of the 2009 report, the cost had risen sharply to $2874 per kW, and in the 2011 report, it had risen sharply again to $4567 per kW. Since then the estimated costs have stabilized once more and the 2015 report estimated the cost for a plant commissioned in 2014 to be $4646 per kW.

The capital cost of nuclear power in the United States makes it one of the most expensive types of plant to build. The cost is similar to off-shore wind and is only exceeded in the 2015 EIA report by coal power with carbon capture and storage and by fuel cells.

Capital costs depend on local conditions and vary from country to country. According to the IEA, the capital cost of nuclear power across the OECD ranges from $2021 per kW in South Korea to $6215 per kW in Hungary. Costs in China were estimated to be $1807 per kW and $2615 per kW for two separate nuclear projects. These variations depend upon factors such as commodity costs and labor costs as well as upon ensuring that projects are completed to schedule.[3]

---

[2]These figures are taken from the US Energy Information Administrations Assumptions to the Annual Energy Outlook, 2001–15.
[3]Projected Costs of Generating Electricity, 2015 Edition, IEA and NEA, 2015.

THE LCOE FROM A NUCLEAR POWER STATION

The LCOE from a nuclear power plant in the United States in 2015 was estimated by Lazard to be between \$97 per MWh and \$136 per MWh.[4] This was broadly similar to the cost of electricity from a coal-fired power plant (without carbon capture and storage) but more expensive than onshore wind power, utility solar power, or electricity from a natural gas combined cycle power plant. It is important to note that the nuclear cost estimate from Lazard does not include the cost of decommissioning which could have a significant effect on the actual cost of power.

As with capital cost, the LCOE varies from country to country. IEA analysis suggests that at a similar discount rate (7%) the cost varies between \$40 per MWh in South Korea and \$101 per MWh in the United Kingdom; it was \$37 per MWh and \$48 per MWh for the two Chinese projects cited above.

Lazard's Levelized Cost of Energy Analysis—Version 9.0, Lazard 2015.

Printed in the United States
By Bookmasters